Introduction to Microstation VBA

saeed murray

Published by saeed murray, 2024.

While every precaution has been taken in the preparation of this book, the publisher assumes no responsibility for errors or omissions, or for damages resulting from the use of the information contained herein.

INTRODUCTION TO MICROSTATION VBA

First edition. October 6, 2024.

Copyright © 2024 saeed murray.

ISBN: 979-8227357779

Written by saeed murray.

Table of Contents

....1
Book Summary..4
1. Introduction to VBA in MicroStation..6
2. Getting Started with VBA in MicroStation 7
3. VBA Code Structure 11
4. Working with Variables... 16
5. Creating VBA Forms 20
6. Adding Levels in MicroStation 29
7. Error Handling in VBA 42
8. Exercise 1: Level Creation App 47
9. Working with Elements 50
10. Manipulating Elements 62
11. Grouping and Ungrouping Elements Using VBA Commands ... 67
12. Retrieving Element IDs in MicroStation VBA 71
13. Calling Subroutines in VBA.. 75
14. Element Locking and Unlocking... 91
15. Conditional Logic in VBA 97
16. Cell Renaming... 103
17. Working with Coordinates 105
18. Enumeration and Scanning Elements... 108
19. Working with Excel 112
20. Working From Excel to Microstation 124
21. Using Excel as a Database 141
and Notepad Using .txt Files 159
23. VBA Calculations in MicroStation... 165
24. Placing Cells in 2D and 3D, and Automating with Excel Data in MicroStation 173
25. Recording Macro Tool... 188
26: Advanced MicroStation UserForm with Excel Integration 197
27. Creating a Micro Station Add-on... 211
Closing Summary 217
About the Author..218

Introduction to Microstation VBA

• • • •

Microstation Connect Edition

• • • •

SAEED MURRAY

© 2024 Saeed Murray.
All rights reserved.
This book, including all text, images, and other content, is protected by copyright law. No part of this publication may be reproduced, stored in a retrieval system, or transmitted in any form or by any means, electronic, mechanical, photocopying, recording, or otherwise, without the prior written permission of the author. For permissions, please contact saeed.murray@gmail.com

Automating MicroStation with **VBA**

Book Summary

This book provides a comprehensive guide to automating tasks in MicroStation using Visual Basic for Applications (VBA). Whether you are new to programming or an experienced MicroStation user looking to enhance your workflows, this book offers a step-by-step approach to mastering VBA in the MicroStation environment.

Starting with the basics of the VBA IDE, variables, and code structure, the book guides you through fundamental programming concepts before diving into MicroStation-specific tasks. You'll learn how to create and manipulate elements, automate drawing processes, and interact with external applications like Excel for data import/export.

Key topics include:

- **Automating Element Creation**: Learn how to programmatically draw lines, circles, arcs, and text, as well as group and manipulate elements.
- **Advanced VBA Projects**: Build practical applications such as the *Geotech Cells App*, an *AI Shape Generator*, and a *Light Coverage Project*, each designed to solve real-world problems.
- **Working with Coordinates**: Import, export, and handle coordinate data with precision, automating tasks that involve complex geometry.
- **Interfacing with Excel & Notepad**: Use Excel and notepad.txt file as a data source or destination, seamlessly exchanging information between MicroStation and Excel or Notepad for reports, analysis, or batch processing.
- **Creating Custom Tools and Add-ons**: Design and implement your own MicroStation add-ons to streamline repetitive tasks and enhance productivity.

THROUGH PRACTICAL EXERCISES, projects, and case studies, this book ensures that you gain hands-on experience with VBA in MicroStation. Whether you're automating simple workflows or tackling complex design problems, this guide will equip you with the tools you need to become proficient in VBA programming for MicroStation.

By the end of the book, you will have developed a solid foundation in VBA, with the ability to create custom applications, automate design processes, and significantly improve your efficiency within MicroStation.

This book is a valuable resource for MicroStation users, CAD professionals, and engineers looking to harness the full power of VBA for automating tasks and enhancing their design workflows.

1.
Introduction to VBA in MicroStation

What is VBA?

- **VBA (Visual Basic for Applications)** is a programming language developed by Microsoft. It is integrated into many Microsoft applications, including MicroStation, Excell Etc.
- VBA allows users to automate repetitive tasks, customize workflows, and create new functionalities within MicroStation, enhancing productivity and efficiency.

Benefits of Using VBA in MicroStation

- **Automation**: Automate repetitive tasks to save time and reduce errors.
- **Customization**: Tailor workflows to meet specific project requirements.
- **Enhanced Functionality**: Create new tools and features that are not available out-of-the-box.
- **Integration**: Seamlessly integrate with other Microsoft applications like Excel for data manipulation and reporting.

2.
Getting Started with VBA in MicroStation

Accessing the VBA Environment:

- Open MicroStation.
- Navigate to **Drawing** > **Utilities** > **Macros** > **VBA Manager**.

Figure 1 - VBA manager

- **Create VBA Project** and save it on you preferred folder.

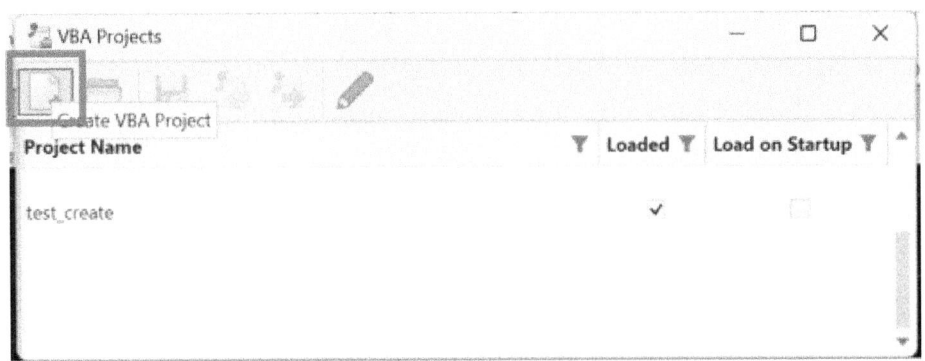

Figure 2 - Create VBA Project

- Select your VBA project and click on the orange pencil on the VBA project window.

Figure 3 - Show VBA IDE

- This will open the VBA IDE (Integrated Development Environment) where you can write and manage your VBA code.

Understanding the VBA IDE:

THE VBA IDE IS A POWERFUL tool that allows you to write, edit, and debug your VBA code. It consists of several key components:

Menus

THE MENU BAR AT THE top of the IDE provides access to various commands and features. Here are some of the main menus:

- **File**: Open, save, and manage your VBA projects.
- **Edit**: Cut, copy, paste, and find/replace code.
- **View**: Toggle different windows and toolbars.
- **Insert**: Add new modules, user forms, and controls.
- **Format**: Adjust the appearance of your code and user forms.
- **Run**: Execute your code and manage breakpoints.
- **Tools**: Access additional tools and options.
- **Debug**: Step through your code and troubleshoot errors.

- **Help**: Access VBA documentation and help resources.

Toolbars

TOOLBARS PROVIDE QUICK access to commonly used commands. You can customize the toolbars to include the buttons you use most frequently. Some standard toolbars include:

- **Standard**: Basic file and editing commands.
- **Debug**: Tools for running and debugging your code.
- **UserForm**: Controls for designing user forms.

Windows

THE IDE CONTAINS SEVERAL windows that help you manage your projects and code:

- **Project Explorer**: Displays all the projects and modules in your VBA environment.
- **Properties Window**: Shows the properties of the selected object, such as a form or control.
- **Code Window**: Where you write and edit your VBA code.
- **Immediate Window**: Allows you to execute VBA commands directly and see the results immediately.
- **Toolbars and Menus**: Provide quick access to various commands and features.

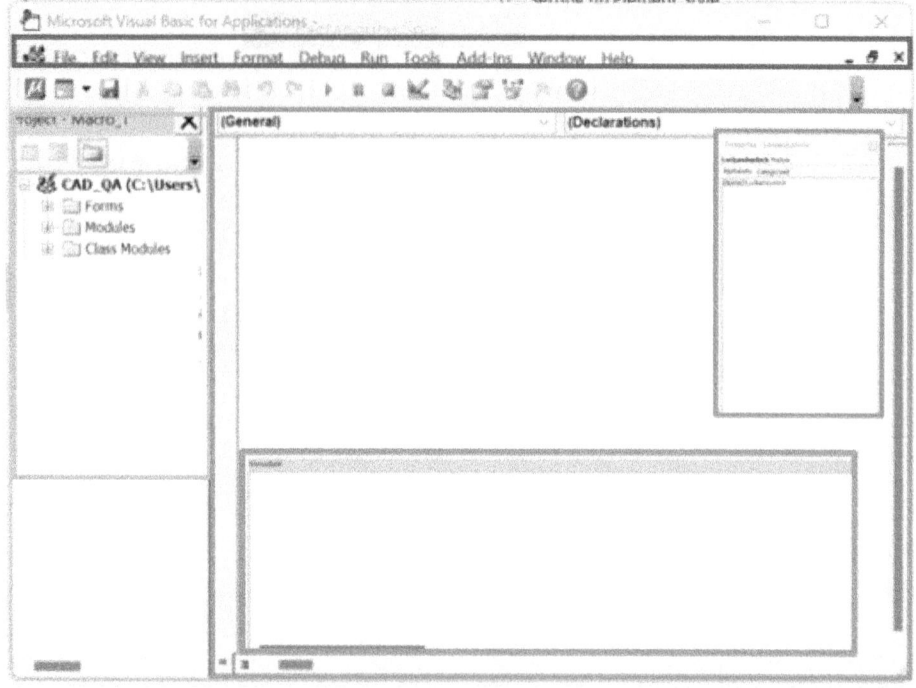

Figure 4 - VBA IDE

Creating Your First Macro:

Go to **Insert > Module** to create a new module.
Write your first VBA code in the code window. For example:

VBA Code 1

SUB HELLOWORLD()
   MsgBox "Hello, World!"
   End Sub

Run the macro by pressing **F5** or using the **Run** menu.

3. VBA Code Structure

Code, Modules and Procedures

Code: A typical code has 4 parts.

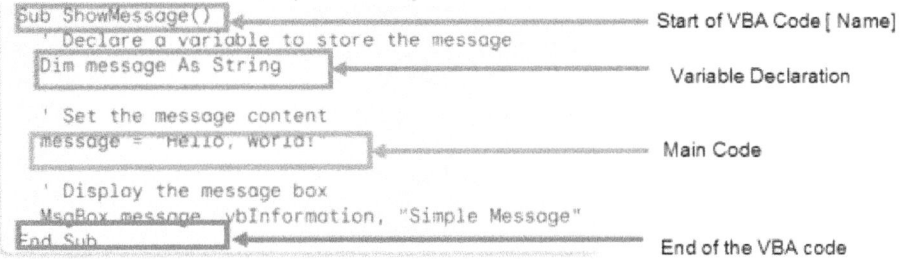

MODULES: Containers for your VBA code. There are two types of modules:

1. Standard Modules:

- Store general procedures and functions.
- Used for code that can be called from anywhere in the project.

2. Class Modules:

- Define objects and their associated procedures.
- Used to create new objects and define their properties, methods, and events.

Procedures: Blocks of code that perform specific tasks. There are two types:

1. Subroutines (Sub):

- Perform actions but don't return values.
- Used for tasks like manipulating objects, displaying messages, or performing calculations without returning a result.

2. Functions:

- Perform actions and return a value.
- Used for calculations or operations that need to return a result to the calling code.

Subroutines and Functions

SUBROUTINES: Defined using the Sub keyword, they do not return a value.

VBA Code 2

```
SUB CREATECIRCLE()
    ' Code to create a circle in MicroStation
    End Sub
```

• • • •

FUNCTIONS: Defined using the Function keyword, they return a value. Below is example that include both a function and a sub routine:

VBA Code 3

```
FUNCTION ADDNUMBERS(a As Integer, b As Integer) As Integer
    AddNumbers = a + b
    End Function
    Sub DisplaySum()
    Dim a As Integer
    Dim b As Integer
    Dim result As Integer
    ' Assign values to a and b
    a = 2
    b = 5
    ' Call the AddNumbers function
    result = AddNumbers(a, b)
    ' Print the result to the Immediate Window
    Debug.Print "The sum of " & a & " and " & b & " is: " & result
```

End Sub

• • • •

IN THIS EXAMPLE, THE AddNumbers **function** takes two integers as arguments and returns their sum.

The DisplaySum **subroutine** assigns values to a and b, calls the AddNumbers function, and prints the result as shown below.

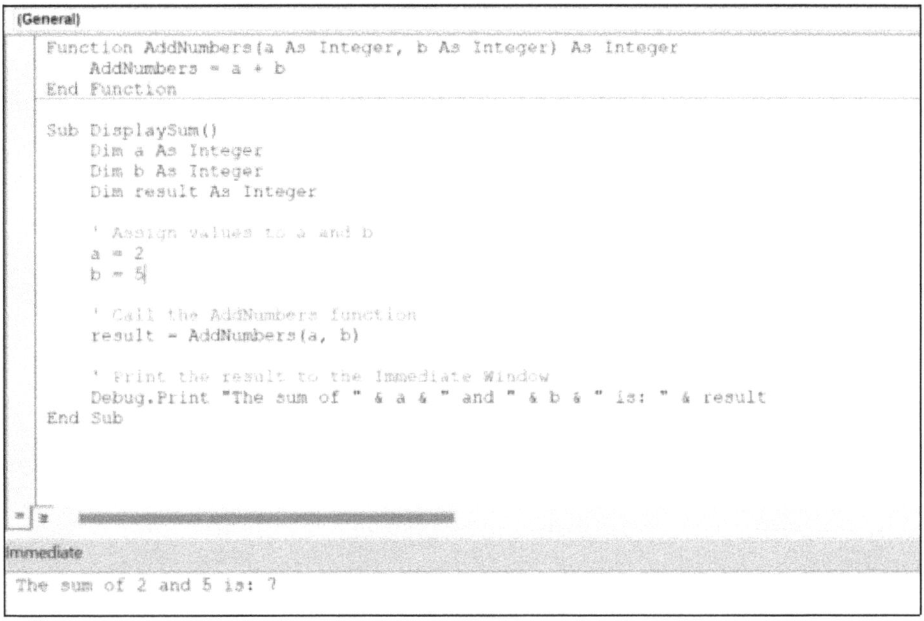

```
(General)
Function AddNumbers(a As Integer, b As Integer) As Integer
    AddNumbers = a + b
End Function

Sub DisplaySum()
    Dim a As Integer
    Dim b As Integer
    Dim result As Integer

    ' Assign values to a and b
    a = 2
    b = 5

    ' Call the AddNumbers function
    result = AddNumbers(a, b)

    ' Print the result to the Immediate Window
    Debug.Print "The sum of " & a & " and " & b & " is: " & result
End Sub
```

```
Immediate
The sum of 2 and 5 is: 7
```

Figure 5 - VBA Code in Microstation

Organizing Your Code

Use modules to logically separate different tasks or functionalities:

IF YOU ARE DEVELOPING a VBA project for a financial application, you might have separate modules for different functionalities such as data input, calculations, and reporting.

- **DataInputModule**: Contains procedures for handling user inputs

and data validation.
- **CalculationModule**: Contains functions and subroutines for performing financial calculations.
- **ReportingModule**: Contains procedures for generating and formatting reports.

VBA Code 4

```
' DATAINPUTMODULE
    Sub GetUserData()
    ' Code to get user data
    End Sub
    ' CalculationModule
    Function CalculateInterest(principal As Double, rate As Double, time As Double) As Double
    CalculateInterest = principal * rate * time
    End Function
    ' ReportingModule
    Sub GenerateReport(principal , rate, time)
    ' Code to generate report
    End Sub
```

Always comment your code for clarity, especially in larger projects:

ADDING COMMENTS TO explain the purpose of procedures, the logic behind complex calculations, or the reason for specific coding decisions. This practice makes your code easier to understand and maintain, especially for others who might work on the project in the future.

VBA Code 5

```
' THIS SUBROUTINE GETS user data from the input form
    Sub GetUserData()
    ' Code to get user data
    End Sub
    ' This function calculates the interest based on principal, rate, and time
    Function CalculateInterest(principal As Double, rate As Double, time As Double) As Double
```

```
' Formula: Interest = Principal * Rate * Time
CalculateInterest = principal * rate * time
End Function
' This subroutine generates a report based on the calculated data
Sub GenerateReport()
' Code to generate report
End Sub
```

4. Working with Variables

Declaring Variables

- **Purpose**: Variables are used to store data that your program will use. They can hold different types of data such as numbers, text, or objects.
- **Syntax**: Declare variables using the Dim keyword followed by the variable name and data type.

VBA Code 6

```
Dim radius As Double
    radius = 5.5
```

Data Types and Scope

Common Data Types in VBA:

Integer: For whole numbers.

VBA Code 7

```
DIM COUNT AS INTEGER
    count = 10
```

Double: For decimal numbers.

VBA Code 8

```
DIM PRICE AS DOUBLE
    price = 19.99
```

***String:** For text.*

VBA Code 9

```
DIM MESSAGE AS STRING
    message = "Hello, World!"
```

***Boolean:** For True/False values.*

VBA Code 10

```
DIM ISACTIVE AS BOOLEAN
    isActive = True
```

***Scope:** Determines where a variable can be accessed. There are three types:*

***Local Scope:** Variables declared inside a procedure are only accessible within that procedure.*

VBA Code 11

```
SUB CALCULATETOTAL()
    Dim total As Double
    total = 100.0
    ' total is only accessible within this subroutine
End Sub
```

***Module Scope:** Variables declared at the top of a module are accessible by all procedures within that module.*

VBA Code 12

```
DIM MODULETOTAL AS Double
    Sub InitializeTotal()
    moduleTotal = 100.0
    End Sub
    Sub DisplayTotal()
```

```
InitializeTotal
MsgBox "Total: " & moduleTotal
End Sub
```

Global Scope: *Variables declared using Public are accessible by all modules in the project.*

VBA Code 13

```
PUBLIC GLOBALTOTAL As Double
    Sub SetGlobalTotal()
    globalTotal = 100.0
    End Sub
```

• • • •

```
SUB SHOWGLOBALTOTAL()
    SetGlobalTotal
    MsgBox "Global Total: " & globalTotal
    End Sub
```

Constants and Enumerations

Constants: *Fixed values that do not change during program execution. Use the Const keyword to declare them.*

VBA Code 14

```
CONST PI AS DOUBLE = 3.14159
```

Enumerations: *A way to assign meaningful names to sets of related constants.*

VBA Code 15

```
ENUM COLOR
    Red = 1
    Green = 2
```

```
Blue = 3
End Enum
Sub DisplayColor()
Dim myColor As Color
myColor = Color.Red
MsgBox "Selected Color: " & myColor
End Sub
```

5.
Creating VBA Forms

Designing a Form

- **Purpose**: VBA forms allow you to create user-friendly interfaces for your automation tasks. They can make your applications more interactive and easier to use.
- **Components**: Forms can contain various controls like buttons, text boxes, combo boxes, etc., to interact with the user.

Steps to Create a Form:

1. Insert a new form by right-clicking in the Project Explorer.
2. Select Insert > UserForm. New form will be shown with the name of UserForm1, unless if the name is taken then it will be names as UserForm2.

Figure 6 - VBA UserForm

1. Insert a new module by right-clicking in the Project Explorer. Select Insert > module.

INTRODUCTION TO MICROSTATION VBA   21

2. Insert the code below to the new module.

VBA Code 15

```
' THIS CODE DEMONSTRATES how to create a simple form
    Sub ShowForm()
    Dim frm As Object
    Set frm = New UserForm1 ' Reference your existing form
    frm.Show
    End Sub
```

• • • •

THE CODE ABOVE SHOULD show the UserForm in you MicroStation window.

Figure 7 - UserForm in MicroStation window

IF YOU HAVE SUCCESSFULLY created a UserForm now let learn about adding controls on the form.

Adding Controls

Step-by-Step Guide to Adding Controls in VBA Using the

Toolbox

Step 1: Open the VBA Editor

1. In MicroStation open the **VBA Editor**.
2. In the VBA Editor, go to **Insert** > **UserForm** to create a new form.

Step 2: Display the Toolbox

1. If the Toolbox isn't already visible, go to **View** > **Toolbox**.
    - The Toolbox contains various controls that you can drag and drop onto your form, such as text boxes, buttons, and labels.

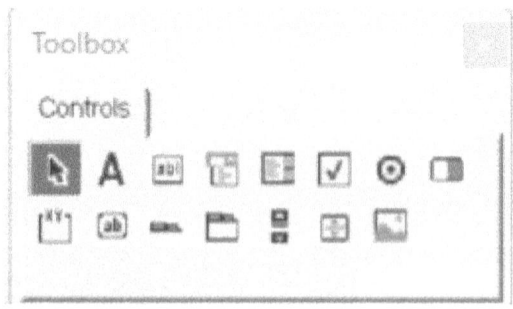

Figure 8 - UserForm Toolbox

Step 3: Add a Text Box

1. **Drag a Text Box from the Toolbox** onto the form:
    - Locate the **TextBox** control in the Toolbox and click on it.
    - Click anywhere on the form to place the text box.
    - Adjust its size and position by dragging the corners.

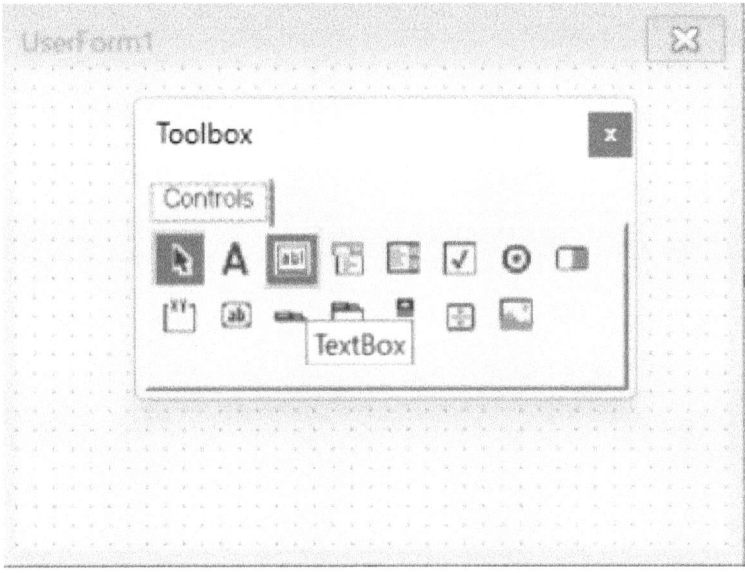

Figure 9 - TextBox

Step 4: Add a Command Button

1. **Drag a Command Button from the Toolbox** onto the form:
    - Find the **CommandButton** control in the Toolbox and click on it.
    - Place it on the form by clicking where you want it to appear.
    - Adjust the size and position if necessary.

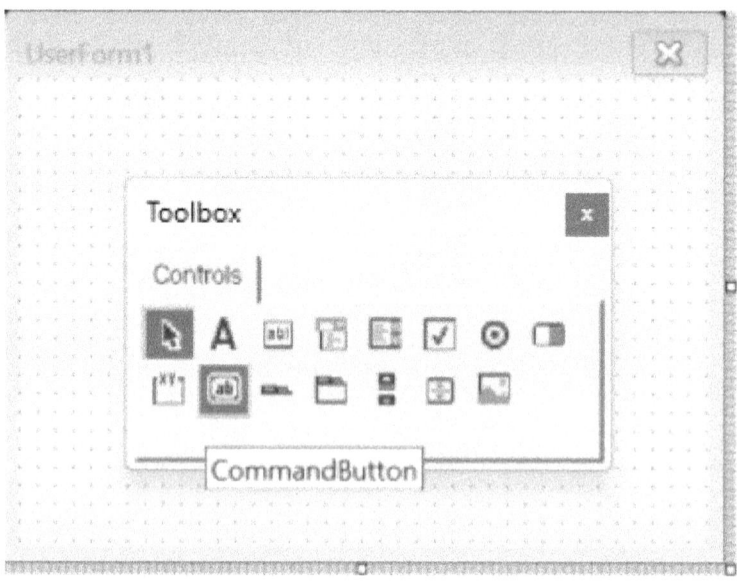

Figure 10 - Command Button

INTRODUCTION TO MICROSTATION VBA

Step 5: Add a Label

1. **Drag a Label from the Toolbox** onto the form:
    - Locate the **Label** control in the Toolbox.
    - Click and drag it onto the form, positioning it where you'd like

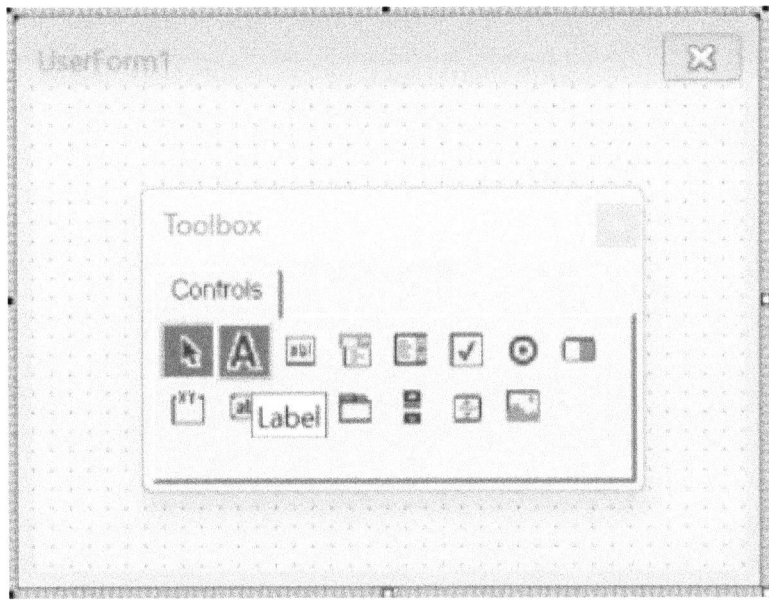

Figure 11 - label

Step 6: Customize the Controls

- You can change properties like **Caption**, **Top**, **Left**, **Width**, etc., for each control by clicking on the control and editing the properties in the **Properties Window**.

Figure 12 - Toolbox Properties

Using VBA UserForm

Step 1: Add UserForm TextBox and Button

- Add **TextBox** and change to the name in properties to *txtName* and add a button to the form and change name to button on properties to *btnshowname* an change caption to Show Name.
- **Double-click the button** to open its click event handler.
- Add code to perform actions when the button is clicked.

VBA Code 16

```
PRIVATE SUB BTNSHOWNAME_Click ()
    MsgBox "Hello " & txtName.Text
    End Sub
```

• • • •

THIS EXAMPLE DISPLAYS a message box with the text the user entered in the text box.

Step 2: Test the Form

- Press F5 or go to **Run** > **Run Sub/UserForm** to display and test your form.
- Interact with the controls to ensure everything is working as expected.

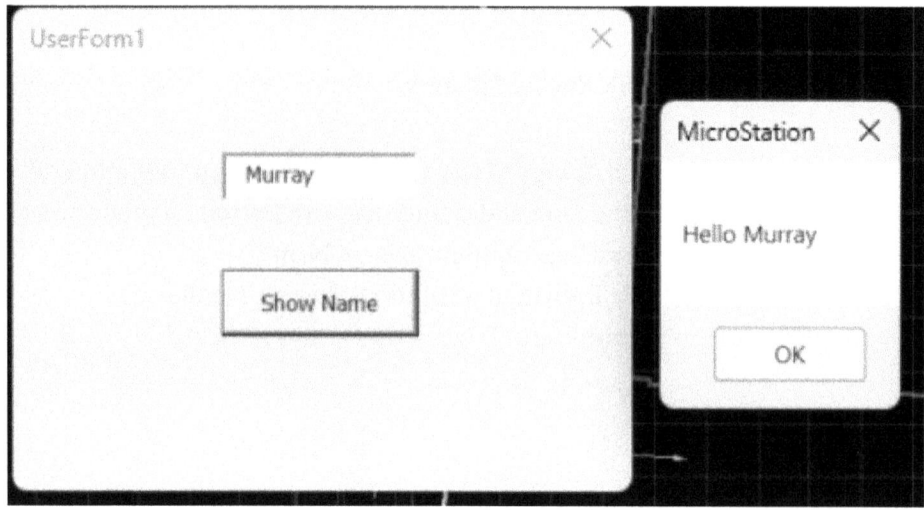

Figure 13 - UserForm exampl

6. Adding Levels in MicroStation

In MicroStation, levels function similarly to layers in other CAD programs. They allow you to organize, manage, and control the appearance of elements in your drawing. With levels, you can change visibility, set different attributes such as color, line style, and weight, and improve the overall structure of your drawing. VBA makes it possible to automate the creation, modification, and management of levels in your drawings.

VBA Code for Adding Levels

USING VBA, YOU CAN add new levels to your MicroStation drawings programmatically. This can be particularly useful when creating complex projects or automating repetitive tasks, such as adding a set of predefined levels to multiple drawings.

Here's how you can create a new level using VBA:

VBA Code 17

```
SUB CREATELEVEL()
    Dim desiredLevelName As String
    desiredLevelName = "M-G55-M_Communications3"
    ActiveDesignFile.AddNewLevel desiredLevelName
    ActiveSettings.Level = ActiveDesignFile.Levels _
    (desiredLevelName)
End Sub
```

Explanation of the Code:

1. **Sub CreateLevel():** This line defines a new subroutine named CreateLevel. Subroutines in VBA are blocks of code that perform a specific task and can be called from other parts of your program.
2. **Dim desiredLevelName As String** : Here, a variable named desiredLevelName is declared. It is of type String, which means it will hold text data. This variable will store the name of the level we want

to create.
3. **desiredLevelName = "M-G55-M_Communications"** : This line assigns the string "M-G55-M_Communications" to the desiredLevelName variable. This is the name of the new level that will be created in the design file.
4. **ActiveDesignFile.AddNewLevel desiredLevelName** : This line calls the AddNewLevel method of the ActiveDesignFile object, passing in desiredLevelName as the parameter. This creates a new level with the specified name in the currently active design file.
5. **ActiveSettings.level = ActiveDesignFile.Levels(desiredLevelName)** : Finally, this line sets the level property of the ActiveSettings object to the newly created level. It accesses the Levels collection of the ActiveDesignFile to find the level by its name (desiredLevelName). This effectively makes the newly created level the active level in the design settings.
6. **End Sub** : This line marks the end of the CreateLevel subroutine.

AN ERROR MAY OCCUR if you run this code if the level name is taken. To deals with this error at this stage add error handler before new level is created
So, the code will look like this.

VBA Code 18

```
SUB CREATELEVELWITHERRORHANDLER()
    Dim desiredLevelName As String
    desiredLevelName = "M-G55-M_Communications"
    On Error Resume Next
    ActiveDesignFile.AddNewLevel desiredLevelName
    ActiveSettings.Level = ActiveDesignFile.Levels _
    (desiredLevelName)
    End Sub
```

With this added line of code, VBA is telling the computer if you can't do a specific line in the code just skip to the next line. In this case it will activate the level named.

It also possible to add levels with descriptions and choose a color as shown in example below.

VBA Code 19

```
SUB CREATENEWLEVELWITHCOLOR()
  ' Create a new level in the active design file
  Dim oLevel As Level
  NewLevelName = "M-G55-M_Camera"
  'Error handler
  On Error Resume Next
  Set oLevel = ActiveDesignFile.AddNewLevel(NewLevelName)
  ' add the level properties
  oLevel.Description = "This is a new level" 'description for
  'the level
  oLevel.ElementColor = 1 ' Set the color (1 = green)
  ' Set the new level as the active level
  ActiveSettings.Level = ActiveDesignFile.Levels(NewLevelName)
End Sub
```

Summary

THE SUBROUTINE WE HAVE seen, creates a new level in the active design file and then sets that level as the active level for further operations.

Modifying Existing Levels

IN ADDITION TO ADDING new levels, you can also modify existing levels. For instance, you may want to change the color, line style, or description of a specific level.

Here's an example of how to modify an existing level:

VBA Code 20

```
SUB MODIFYEXISTINGLEVEL()
  Dim oLevel As Level
  ' Retrieve an existing level by its name
  ExistingLevelName = "M-G55-M_Communications"
  Set oLevel = ActiveDesignFile.Levels(ExistingLevelName)
  ' Modify the level properties
  oLevel.Description = "Updated level description"
  oLevel.ElementColor = 2 ' Change the color (2 = green)
```

End Sub

Explanation of the Code:

1. ActiveDesignFile.Levels("ExistingLevelName"): This retrieves an existing level by its name.
2. The properties (Description, Color, LineStyle) of the level can be updated in the same way as for a newly created level.

Additional Level Attributes

HERE ARE SOME OTHER useful properties and methods you can use when working with levels in VBA:

- **Visibility Control:** You can control whether a level is visible or hidden using the IsVisible property.

VBA Code 21

OLEVEL.ISDISPLAYED = False ' Hide the level

- **Locking Levels:** Levels can be locked to prevent changes or accidental modifications.

VBA Code 22

OLEVEL.ISLOCKED = TRUE ' Lock the level

- **Line Weight:** You can specify the line weight for elements on the level.

VBA Code 23

OLEVEL.LINEWEIGHT = 3 ' Set the line weight to 3

Practical Example: Adding Multiple Levels

YOU CAN ALSO CREATE multiple levels programmatically, especially useful when working with large projects that require several predefined levels. In This Example we will look at using a For loop.

Structure of a For Loop

THE BASIC SYNTAX OF a For loop in VBA (and many other languages) looks like this:

VBA Code 24

```
FOR COUNTER = START To end [Step step]
    ' Code to execute in each iteration
    Next counter
```

Components of the For Loop:

1. COUNTER:

> This is a variable that keeps track of the current iteration of the loop. It is typically an integer.

2. Start:

> This is the initial value of the counter when the loop begins.

3. End:

> This is the final value of the counter. The loop will continue to execute as long as the counter is less than or equal to this value.

4. Step (Optional):

> This is an optional parameter that defines the increment or decrement for the counter. If not specified, it defaults to 1. For example, Step 2 would increase the counter by 2 each iteration.

5. Code Block:

This is the block of code that will execute on each iteration of the loop.

6. Next:

This keyword marks the end of the loop. It increments the counter and checks the loop condition for the next iteration.

How a For Loop Works

1. **Initialization**: The counter is initialized to the start value.
2. **Condition Check**: Before each iteration, the loop checks if the counter has reached the end value.
3. **Execution**: If the condition is true, the code block within the loop executes.
4. **Increment**: After executing the code block, the counter is updated (incremented or decremented based on the Step value).
5. **Repeat**: The process repeats until the counter exceeds the end value.

EXAMPLE OF A FOR LOOP

Here's a simple example of a For loop in VBA:

VBA Code 25

```
SUB COUNTTOTEN()
    Dim i As Integer
    For i = 1 To 10
    Debug.Print i ' This will print numbers 1 to 10 in the
    'Immediate Window
    Next i
End Sub
```

In this example:

- The loop starts with i initialized to 1.
- It prints the value of i in the Immediate Window.

- After printing, i is incremented by 1 until it reaches 10, at which point the loop stops.

Benefits of Using a For Loop

1. **Automation of Repetitive Tasks**: Loops allow you to automate tasks that need to be repeated multiple times, reducing manual effort and errors.
2. **Dynamic Iteration**: You can easily adjust the start, end, and step values to control how many times the loop runs.
3. **Improved Code Readability**: Using a loop makes your code cleaner and easier to understand, especially when dealing with repetitive operations.
4. **Flexibility**: You can easily adapt a loop for different scenarios by changing the range or step size.
5. **Efficient Resource Use**: Instead of writing the same code multiple times, a loop allows you to reuse the code block, which can improve performance, especially in large projects.

THE FOR LOOP IS A POWERFUL and essential concept in programming that simplifies repetitive tasks, enhances code organization, and improves overall efficiency. By mastering loops, you can automate a wide range of processes and manage data more effectively in your applications. Now let's try to use for loop to automate multiple levels creation in Microstation.

VBA Code 26

```
SUB CREATEMULTIPLELEVELS()
    Dim levelsArray As Variant
    levelsArray = Array("Walls", "Windows", "Doors", "Furniture")
    Dim i As Integer
    For i = LBound(levelsArray) To UBound(levelsArray)
    Dim oLevel As Level
    Set oLevel = ActiveDesignFile.AddNewLevel(levelsArray(i))
    ' Set attributes for each level
    oLevel.Description = "Level for " & levelsArray(i)
    oLevel.ElementColor = i + 1 ' Assign a unique color to
    'each level
    Next i
    End Sub
```

Explanation of the Code:

THE LEVELSARRAY CONTAINS the names of the levels you want to create. A **For loop** iterates through the array, creating a new level for each name and assigning attributes such as description and color.

Breakdown of the Code:

1. SUBROUTINE DECLARATION:

- Sub CreateMultipleLevels() starts the declaration of a subroutine called CreateMultipleLevels.

2. Array Declaration:

- Dim levelsArray As Variant: This declares a variable levelsArray that can hold an array of values.
- levelsArray = Array("Walls", "Windows", "Doors", "Furniture"): This initializes levelsArray with four predefined level names.

3. Loop Declaration:

- Dim i As Integer: This declares a loop counter variable i.
- For i = LBound(levelsArray) To UBound(levelsArray): This starts a For loop that will iterate from the lower bound (LBound) to the upper bound (UBound) of the levelsArray. In this case, it will loop through the indices 0 to 3.

4. Creating New Levels:

- Dim oLevel As Level: This declares a variable oLevel of type Level, which will represent each new level created.
- Set oLevel = ActiveDesignFile.AddNewLevel(levelsArray(i)): Inside the loop, this line creates a new level in the active design file using the current element of the levelsArray.

5. Setting Attributes:

- oLevel.Description = "Level for " & levelsArray(i): This sets a description for the new level, concatenating the string "Level for " with the current level name.
- oLevel.ElementColor = i + 1: This assigns a unique color to each level based on its index in the array.

6. End of Loop:

- Next i: This signifies the end of the For loop, which then increments i and repeats until all elements in levelsArray have been processed.

Another type of loop is the **Do while loop** . The Do While loop is a control structure in programming that allows you to execute a block of code repeatedly as long as a specified condition is true. It is particularly useful for scenarios where the number of iterations is not known beforehand but depends on a certain condition being met.

Structure of a Do While Loop

THE BASIC SYNTAX OF a Do While loop in VBA is as follows:

VBA Code 27

DO WHILE CONDITION
    ' Code to execute as long as condition is True
Loop

Components of the Do While Loop:

1. CONDITION:

- This is a boolean expression that is evaluated before each iteration of the loop. If it evaluates to True, the loop executes; if it evaluates to False, the loop exits.

2. Code Block:

- This is the block of code that will be executed as long as the condition is True.

3. Loop:

- This keyword marks the end of the loop block. Once the condition evaluates to False, the execution continues after the loop.

How a Do While Loop Works

1. CONDITION CHECK: Before executing the code block, the loop checks the condition.
    2. Execution: If the condition is True, the code block executes.
    3. Repeat: After executing the code, the condition is checked again. If it's still True, the loop repeats.

4. Exit: When the condition evaluates to False, the loop terminates, and execution continues with the next line of code following the loop.

Example of a Do While Loop

HERE'S A SIMPLE EXAMPLE of a Do While loop in VBA:

VBA Code 28

```
SUB COUNTTOTENLOOP()
    Dim count As Integer
    count = 1
    Do While count <= 10
    Debug.Print "Count: " & count ' Print the current count
    count = count + 1 ' Increment the count
    Loop
    Debug.Print "Finished counting!" ' Indicate that counting is
    'done
    End Sub
```

Explanation of the Example:

1. INITIALIZATION:

- The variable count is initialized to 1.

2. Condition Check:

- The loop checks if count is less than or equal to 10. As long as this condition is true, the loop continues to execute.

3. Execution:

- Inside the loop, the current value of count is printed to the Immediate Window using Debug.Print.

4. Increment:

- After printing, count is incremented by 1 (count = count + 1).

5. Repeat:

- The condition is checked again after each iteration. When count exceeds 10, the loop stops.

6. Exit:

- After exiting the loop, a message indicating that counting is finished is printed.

Benefits of Using a Do While Loop

1. FLEXIBILITY:

- The Do While loop is very flexible and can be used in scenarios where the number of iterations is not predetermined, making it ideal for tasks like reading data until the end of a file or processing user input until a specific condition is met.

2. Dynamic Control:

- It allows for dynamic control over the loop execution, as the loop can continue until an external condition changes (e.g., user input, file state).

3. Simplicity:

- The structure of the Do While loop is straightforward, making it easy to understand and implement.

For loop and The Do While loop are powerful and versatile construct for handling repetitive tasks in VBA and other programming languages. By mastering these loops, you can create efficient and flexible code that adapts to various conditions and scenarios.

Practical Example in Automation

IN A PRACTICAL SCENARIO, this type of automation can be particularly useful in architectural design, engineering projects, or any application where multiple layers or categories need to be organized systematically. For instance, if you frequently create design files with the same structure (like walls, windows, and furniture), this automated approach saves time and ensures accuracy across projects.

Conclusion

MANAGING LEVELS THROUGH VBA allows you to streamline your workflows and maintain consistency across multiple drawings. Whether you're creating new levels, modifying existing ones, or automating the setup of complex projects, using levels effectively ensures your drawings are organized, clear, and easy to manage. Through the use of the provided VBA code examples, you can easily implement and customize the management of levels within your own MicroStation projects.

7.
Error Handling in VBA

Error handling is crucial in any programming language, including VBA, to ensure that your code can handle unexpected situations gracefully. As a programmer you will come across with these and its good practice to know how to handle them. We did saw some application on our previous chapter and let's delve deeper into the different error types in VBA and provide examples for each.

Error Types in VBA

Syntax Errors

DESCRIPTION: Syntax errors occur when the code structure is incorrect. The VBA compiler cannot interpret the code, leading to compilation failure.

VBA Code 29

```
SUB SYNTAXERROREXAMPLE()
    Dim x As Integer
    x = 10
    If x > 5 Then ' Missing Then
    MsgBox "x is greater than 5"
    End Sub
```

In the example above we have *"if"* but we are missing Endif. To fix this *"End if"* statement should be added as shown below.

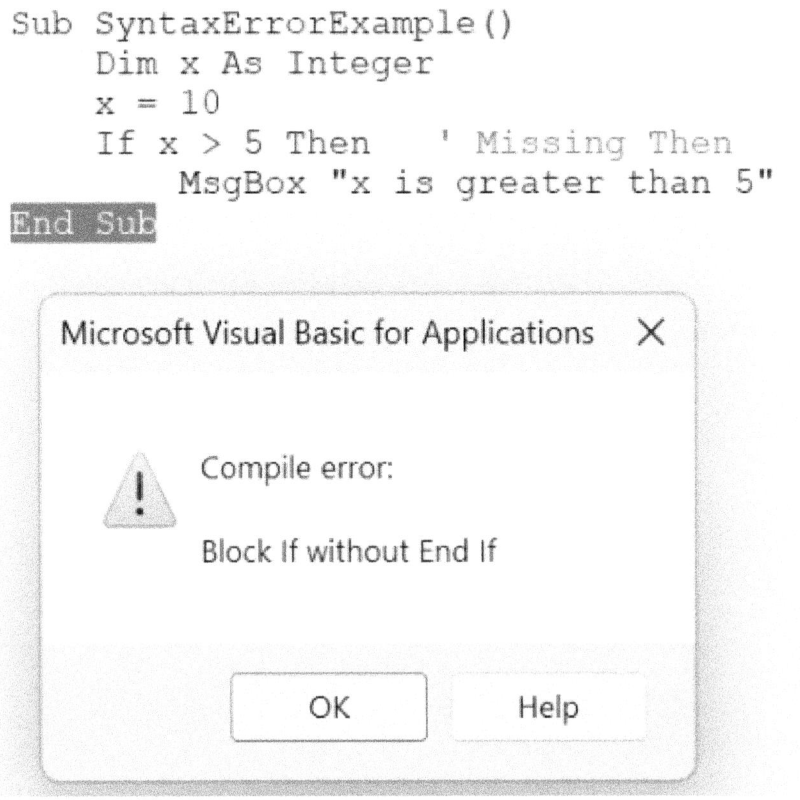

INTRODUCTION TO MICROSTATION VBA 43

```
Sub SyntaxErrorExample()
    Dim x As Integer
    x = 10
    If x > 5 Then    ' Missing Then
        MsgBox "x is greater than 5"
End Sub
```

Microsoft Visual Basic for Applications

Compile error:

Block If without End If

[OK]   [Help]

Figure 14 - Error message

Example: Corrected code

VBA Code 30

SUB SYNTAXERROREXAMPLE()
    Dim x As Integer
    x = 10
    If x > 5 Then
    MsgBox "x is greater than 5"
    End If
    End Sub

Runtime Errors

DESCRIPTION: Runtime errors occur during code execution due to various reasons such as invalid operations (e.g., dividing by zero, accessing an out-of-bounds array element, or using an object that hasn't been set).

VBA Code 31

```
SUB RUNTIMEERROREXAMPLE()
    Dim x As Integer
    Dim y As Integer
    y = 0
    x = 10 / y ' Division by zero will cause a runtime error
End Sub
```

> To fix this, error handling to manage unexpected scenarios, as shown below

VBA Code 32

```
SUB RUNTIMEERROREXAMPLE()
    On Error GoTo ErrorHandler
    Dim x As Integer
    Dim y As Integer
    y = 0
    x = 10 / y
    Exit Sub
ErrorHandler:
    MsgBox "Error: Division by zero!"
End Sub
```

Logical Errors

DESCRIPTION: Logical errors occur when the code runs without any syntax or runtime errors but produces incorrect results due to flawed logic.

Example:

VBA Code 33

```
SUB LOGICALERROREXAMPLE()
    Dim total As Integer
    Dim price As Integer
    price = 100
    total = price * 0.1 ' Incorrect calculation for total
    MsgBox "Total price is: " & total ' This will show 10, not 110
End Sub
```

To fix this, logic must be corrected to show 110, as shown below

VBA Code 34

```
SUB LOGICALERROREXAMPLE()
    Dim total As Integer
    Dim price As Integer
    price = 100
    total = price + (price * 0.1) ' Correct calculation for total
    MsgBox "Total price is: " & total ' This will now show 110
End Sub
```

Implementing Error Handling

TO EFFECTIVELY MANAGE errors in VBA, you can use the On Error statement. Here's how to implement it:

- **On Error Resume Next**: This tells VBA to continue executing the next line of code even if an error occurs.
- **On Error GoTo [Label]**: This redirects the flow of execution to a

specific line (label) in case of an error.

Example of a robust error-handling routine:

VBA Code 35

```
SUB ERRORHANDLINGEXAMPLE()
    On Error GoTo ErrorHandler ' Redirects to ErrorHandler if an error occurs
    Dim x As Integer
    Dim y As Integer
    y = 0
    x = 10 / y ' This will cause a runtime error
    Exit Sub ' Ensures we skip the error handling when no error
    'occurs
    ErrorHandler:
    MsgBox "An error occurred: " & Err.Description ' Display the
    'error message
    End Sub
```

Summary

UNDERSTANDING ERROR types—syntax, runtime, and logical errors—and implementing effective error handling is essential for writing robust VBA code. By anticipating potential issues and managing them gracefully, you can enhance the reliability and user experience of your applications

8.
Exercise 1: Level Creation App

Now it's about time to put everything we have so far and make our own VBA software. In this exercise, you will create a simple VBA application that allows users to create a new level in MicroStation. Users will input the level name, color, and line style through a user-friendly form.

Steps to Build the Level Creation App:

1. DESIGN THE FORM:
- Create a form with the following input fields:

  1. **Level Name**: A text box for the user to enter the name of the new level.
  2. **Color**: A numeric input for the level color.
  3. **Line Weight**: A numeric input for the line thickness.

- Include a "Create Level" button to trigger the level creation process.

2. Write the Code:
- Use the *ActiveDesignFile. AddNewLevel* method to create a new level based on the user's input.

3. Handle User Input:
- Add validation to ensure that the user provides valid inputs. For example, check that the level name is not blank, and that the color and line style are valid numeric values.

4. Run the App:
- Test the application to ensure it successfully creates levels dynamically within MicroStation. Make any necessary adjustments based on user feedback.

Solution:

THE FORM SHOULD LOOK like what is shown below.

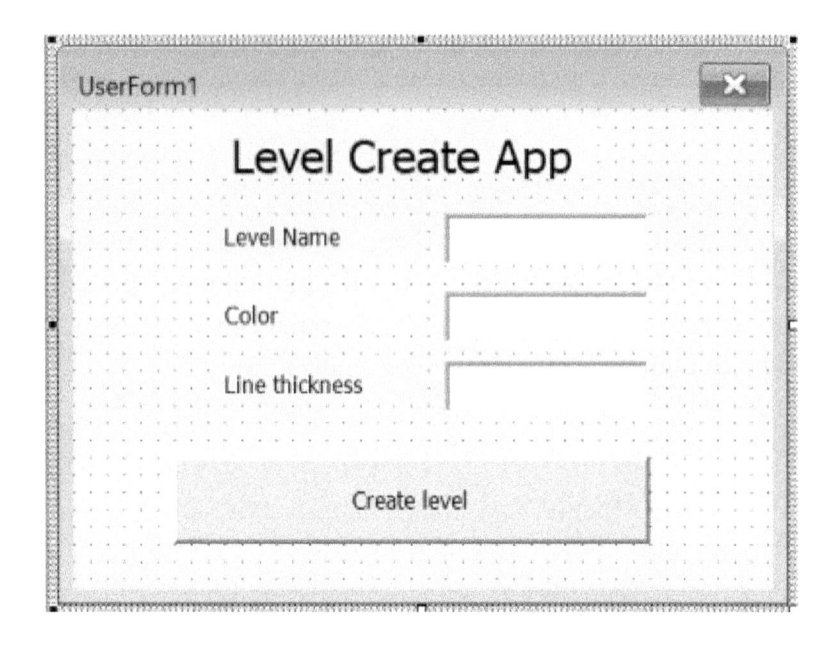

Figure 15 - Level Create UserForm

HERE'S AN EXAMPLE OF the VBA code that you can use for the form's button click event. Double click Create level button in the UserForm and add the following code.

VBA Code 36

```
PRIVATE SUB BTNCREATE_Click()
    Dim oLevel As Level
    ' Validate user input
    If Trim(txtLevelName.value) = "" Then
    MsgBox "Please enter a level name.", vbExclamation
    Exit Sub
    End If
    ' Create a new level
    Set oLevel = ActiveDesignFile.AddNewLevel(txtLevelName.value)
    oLevel.ElementColor = CInt(txtColor.value)
    oLevel.ElementLineWeight = CInt(txtLineStyle.value)
    ActiveSettings.Level = ActiveDesignFile.Levels _
    (txtLevelName.value)
    ' Notify the user
```

MsgBox "Level Created!"
End Sub

For this example, we didn't have to any module for coding. Everything is managed on the UserForm. Once the code is entered, go to UserForm , run the Run the app and try to enter some levels , color and line thickness using the form.

9.
Working with Elements

Drawing Lines with VBA

In MicroStation, lines are simple straight elements defined by two points in a three-dimensional space. Drawing lines programmatically allows for precise control over their placement and attributes.

Example of Drawing a Line:

THE FOLLOWING CODE demonstrates how to draw a line between two specified points:

VBA Code 37

```
SUB DRAWLINE()
    Dim oLine As LineElement
    Dim startPoint As Point3d
    Dim endPoint As Point3d
    startPoint = Point3dFromXYZ(0, 0, 0) ' Starting point at the
    'origin
    endPoint = Point3dFromXYZ(100, 100, 0) ' Ending point _
    'at (100, 100, 0)
    Set oLine = CreateLineElement2(Nothing, startPoint, endPoint) ' Create the line element
    ActiveModelReference.AddElement oLine ' Add the line to the 'active model
End Sub
```

In this example:

- **Point3dFromXYZ**: This function creates a point in 3D space.
- **CreateLineElement2**: This method constructs the line element.
- **ActiveModelReference.AddElement**: This method adds the newly created line to the active model in MicroStation.

String Line in VBA

THE FOLLOWING SUBROUTINE demonstrates how to create a string line in MicroStation using predefined coordinates stored directly in the code. This approach allows for quick and efficient drawing without the need for external file handling.

Example of String Line:

THE FOLLOWING CODE demonstrates how to draw a String line between several given points:

VBA Code 38

```
SUB DRAWSTRING()
    Dim point As Point3d
    Dim coordinates As Variant
    Dim i As Long
    ' Define the coordinates as an array
    coordinates = Array( _
    Array(79350.418, 35116.831, 91.431), _
    Array(79354.426, 35119.814, 91.483), _
    Array(79358.461, 35122.768, 91.506), _
    Array(79362.646, 35125.792, 91.517) _
    )
    ' Start the line string command
    CadInputQueue.SendKeyin "place lstring point"
    ' Loop through the predefined coordinates
    For i = LBound(coordinates) To UBound(coordinates)
    ' Parse coordinates
    point.x = CDbl(coordinates(i)(0))
    point.y = CDbl(coordinates(i)(1))
    point.Z = CDbl(coordinates(i)(2))
    ' Send the data point to the current command
    CadInputQueue.SendDataPoint point, 1
    Next i
    ' Send a reset to the current command
    CadInputQueue.SendReset
    ' End the command
    CommandState.StartDefaultCommand
```

End Sub

Summary

THIS SUBROUTINE EFFICIENTLY creates a string line in MicroStation using hardcoded coordinates. By leveraging loops and point manipulation, it provides a straightforward way to automate drawing tasks, enhancing productivity for users who work with geometric data.

Creating Circles Programmatically

CIRCLES IN MICROSTATION are defined by a center point and a radius. Creating circles programmatically allows for easy modification and precise placement.

Example of Drawing a Circle:

THE FOLLOWING CODE illustrates how to draw a circle with a given center and radius:

VBA Code 39

```
SUB DRAWCIRCLE()
    Dim oCircle As EllipseElement
    Dim centerPoint As Point3d
    centerPoint = Point3dFromXYZ(50, 50, 0) ' Center point at (50, '50, 0)
    Set oCircle = CreateEllipseElement2(Nothing, _
    centerPoint, 25, 25, Matrix3dIdentity) ' Create the circle
    ActiveModelReference.AddElement oCircle ' Add the circle
    End Sub
```

In this example:

- **CreateEllipseElement2**: This method is used to create an ellipse or circle based on the specified center and radii.
- **Matrix3dIdentity**: This parameter represents the transformation matrix, which is set to identity in this case, indicating no transformation.
-

Working with Arcs and Curves

ARCS CAN BE CREATED similarly to circles but require additional parameters such as start and end angles. This allows for greater flexibility in design.

Example of Creating an Arc:

THE FOLLOWING CODE shows how to create an arc defined by a center point, radii, and angles:

VBA Code 40

```
SUB DRAWARC1()
    Dim MyArc As ArcElement
    Dim CenPt As Point3d
    Dim RotMatrix As Matrix3d
    ' Convert 45 degrees and 90 degrees directly to radians
    Set MyArc = Application.CreateArcElement2 _
    (Nothing, CenPt, 1.5, 1.5, RotMatrix, Radians(45), Radians(90))
    Application.ActiveModelReference.AddElement MyArc
End Sub
```

In this example:

- **CreateArcElement1**: This method creates an arc element, where the parameters include the center, radii, start angle, and end angle.

Working with Text in MicroStation

TEXT IS AN ESSENTIAL element in any design and drafting environment. In MicroStation, you can automate the creation and formatting of text using VBA, which allows you to add annotations, labels, and other textual information programmatically. This section covers how to add text to your drawing, format it, and ensure that it is positioned correctly for readability and visual clarity.

Adding Text Using VBA

YOU CAN CREATE AND place text elements in MicroStation using VBA. The text can be customized for size, font, and style. Here's an example of how to create and place text in a design file:

VBA Code 41

```
SUB ADDTEXT()
    Dim oText As TextElement
    Dim origin As Point3d
    ' Define the insertion point for the text
    origin = Point3dFromXYZ(50, 50, 0)
    ' Create a text element with the specified origin and identity 'matrix
    Set oText = CreateTextElement1 _
    (Nothing, "Hello, MicroStation!", origin, Matrix3dIdentity)
    ' Set the text style (height and width of the text)
    oText.TextStyle.Height = 2
    oText.TextStyle.Width = 2
    ' Add the text element to the active model
    ActiveModelReference.AddElement oText
End Sub
```

• • • •

IN THIS EXAMPLE:

- **CreateTextElement1** is used to create a TextElement. This method takes parameters such as the text content ("Hello, MicroStation!"), the insertion point (origin), and the transformation matrix (Matrix3dIdentity).
- The TextStyle properties (Height and Width) are used to define the size of the text.
- Finally, the text is added to the active model using AddElement.

INTRODUCTION TO MICROSTATION VBA

Formatting and Positioning Text

ONCE A TEXT ELEMENT has been created, you can further modify its appearance by changing its style and position. MicroStation allows you to customize the height, width, justification, and other formatting options through the TextStyle object. Here's how to modify the text style and positioning:

VBA Code 42

```
SUB FORMATTEXT()
    Dim oText As TextElement
    Dim origin As Point3d
    ' Define the origin for the text
    origin = Point3dFromXYZ(100, 50, 0)
    ' Create a new text element
    Set oText = CreateTextElement1(Nothing, "Formatted Text", _
    origin, Matrix3dIdentity)
    ' Set text height and width
    oText.TextStyle.Height = 3
    oText.TextStyle.Width = 2
    ' Set text justification to center-center
    oText.TextStyle.Justification = _
    msdTextJustificationCenterCenter
    ' Add the formatted text element to the active model
    ActiveModelReference.AddElement oText
End Sub
```

In this example:

- **Text Height and Width**: oText.TextStyle.Height and oText.TextStyle.Width are used to set the size of the text.
- **Text Justification**: The Justification property is set to msdTextJustificationCenterCenter, which centers the text both horizontally and vertically.
- By customizing these properties, you ensure that your text is properly aligned and scaled according to the drawing context.

Best Practices for Handling Text

WHEN ADDING AND FORMATTING text in MicroStation, it's essential to follow a few best practices to ensure that the text is clear and well-positioned:

1. **Consider the Scale**: Depending on the size and complexity of your drawing, ensure that the text height and width are appropriate. Text that is too small or too large can be hard to read and may overlap with other elements.
2. **Use Appropriate Justification**: When placing text in different contexts (e.g., annotations, titles, labels), adjust the text justification to ensure that it aligns correctly with surrounding elements.
3. **Choose Readable Colors**: Use text colors that contrast well with the background or surrounding elements. In dark areas, use lighter text colors, and in light areas, use darker text.
4. **Layer Control**: Place text elements on a separate layer to control their visibility independently from other elements. This helps when generating views or prints, as you can toggle text layers on or off as needed.
5. **Positioning in Complex Drawings**: In large or detailed drawings, carefully position text to avoid overcrowding. Space text elements apart and avoid placing text directly on top of other design elements, as this can reduce readability.

By following these guidelines and examples, you can efficiently manage text creation and formatting in MicroStation using VBA, ensuring that your drawings remain clear, legible, and professional.

Creating Cells in MicroStation

CELLS ARE AN ESSENTIAL feature in MicroStation that allow users to create reusable symbols or grouped elements. Now we know how to create elements in Microstation using VBA, it is time to do them together. This chapter explains what cells are, how to create both 2D and 3D cells programmatically using VBA and offers best practices for working with cells in MicroStation.

Creating 2D and 3D Cells Programmatically

YOU CAN PROGRAMMATICALLY create 2D or 3D cells in MicroStation using VBA by defining the elements you want to include in the cell and then grouping them into a cell.

Creating a 2D Cell

THE FOLLOWING EXAMPLE demonstrates how to create a simple 2D cell made up of two-line elements, a circle and a Text.

VBA Code 43

```
SUB CREATE2DCELL()
    Dim oCell As CellElement
    Dim oLine As LineElement
    Dim oCircle As EllipseElement
    Dim oText As TextElement
    Dim elements(0 To 3) As Element
    Dim Cell_Name As String
    Cell_Name = "My2DCell"
    ' Create the first line
    Set oLine = CreateLineElement2(Nothing, _
    Point3dFromXYZ(-5, -5, 0), Point3dFromXYZ(5, 5, 0))
    Set elements(0) = oLine ' Add the line to the elements array
    ' Create the second line
    Set oLine = CreateLineElement2(Nothing, _
    Point3dFromXYZ(5, -5, 0), Point3dFromXYZ(-5, 5, 0))
    Set elements(1) = oLine ' Add the line to the elements array
    ' Create a new text element
    Set oText = CreateTextElement1(Nothing, _
    "My Cell", Point3dFromXYZ(0, -10, 0), Matrix3dIdentity)
    ' Format the text element
    oText.TextStyle.height = 2.5 ' Set text height
    oText.TextStyle.width = 2.5 ' Set text width
    oText.TextStyle.Justification = _
    msdTextJustificationCenterCenter
    ' Set text justification to center
    Set elements(2) = oText ' Add the text to the elements array
    ' Create the circle
```

```
Set oCircle = CreateEllipseElement2(Nothing, _
Point3dFromXYZ(0, 0, 0), 5, 5, Matrix3dIdentity)
Set elements(3) = oCircle
' Add the circle to the elements array
' Create a 2D cell from the defined elements
Set oCell = CreateCellElement1(Cell_Name, _
elements, Point3dFromXYZ(0, 0, 0))
' Add the cell to the active model reference
ActiveModelReference.AddElement oCell
End Sub
```

In this example:

- **Lines as Elements**: We create two-line elements and store them in an array.
- **Circle as Ellipse Element**: A circle was created a part of the cell.
- **Text as Text Element**: A text was also added and formatted to the cell.
- **CreateCellElement1**: This method creates a cell named "My2DCell" that includes the two lines and is positioned at the origin (0, 0, 0).
- **ActiveModelReference.AddElement**: Adds the newly created cell to the current model for reuse.

Creating 3D Cells

TO CREATE A 3D CELL, you simply need to define 3D elements (such as 3D lines, arcs, or surfaces) and include them in the cell. The process is similar to creating 2D cells, but the Z-coordinates will define the depth.

Example of creating a 3D cell:

VBA Code 44

```
SUB CREATE3DCELL()
    Dim oCell As CellElement
    Dim oLine As LineElement
    Dim oCircle As EllipseElement
    Dim elements(0 To 1) As Element
    ' Create a 3D line
```

```
Set oLine = CreateLineElement2(Nothing, _
Point3dFromXYZ(0, 0, 0), Point3dFromXYZ(10, 10, 10))
Set elements(0) = oLine ' Add the line to the elements array
' Create a 3D circle (treated as an ellipse)
Set oCircle = CreateEllipseElement2 _
(Nothing, Point3dFromXYZ(5, 5, 5), 2, 2, Matrix3dIdentity)
Set elements(1) = oCircle
' Add the circle to the elements array
' Create a 3D cell
Set oCell = CreateCellElement1 _
("My3DCell", elements, Point3dFromXYZ(0, 0, 0))
' Add the cell to the active model reference
ActiveModelReference.AddElement oCell
End Sub
```

In this 3D example:

- **3D Line**: The first line is defined using 3D coordinates.
- **3D Circle**: An ellipse element is created to represent a circle, with a radius of 2 units in the X and Y directions, centered at (5, 5, 5) in 3D space.
- **Cell Creation**: The CreateCellElement1 method is used to group these elements into a 3D cell named "My3DCell", which is then added to the model.

Best Practices for Creating Cells

WHEN WORKING WITH CELLS in MicroStation, it's important to follow certain best practices to ensure efficiency, clarity, and ease of reuse.

1. Descriptive Cell Naming:

- Always assign descriptive names to cells. This helps in identifying the purpose of the cell when it is reused across different models and drawings.
- Use a naming convention that makes the purpose of the cell immediately clear (e.g., "DoorSymbol", "WindowCell_1x2", "FurnitureTable").

2. Define the Insertion Point Carefully:

- The insertion point defines how the cell is positioned when placed in the model. Choose this point based on how you expect the cell to be placed in the drawing.
- For example, when creating a window or door cell, you may want the insertion point at the corner or center of the object to make placement easier.

3. Group Relevant Elements Only:

- Ensure that all elements in the cell are logically grouped together. Avoid adding unnecessary elements that are not part of the cell's intended function.
- For instance, in architectural drawings, group only the components of a door or window that are relevant to the design, such as the frame, glass, and sill.

4. Use Layers and Levels:

- When creating cells, make sure the elements are placed on appropriate layers or levels. This allows for better visibility control when working in complex models.
- Cells should be structured in such a way that turning off a layer (e.g., for electrical wiring or furniture) does not affect the rest of the design unnecessarily.

5. Reuse of Standard Components:

- Cells are ideal for standard components such as symbols, furniture, or repetitive design elements. Use cells to maintain consistency across multiple models.
- By using a standardized set of cells, you ensure that all instances of a design element (e.g., a specific door type) remain consistent across the project.

6. Easy Modification:

- Cells can be modified if needed. Ensure that when you create a cell, it remains easy to adjust or update in the future without having to rebuild it from scratch.

By following these guidelines and using VBA to automate the creation of 2D and 3D cells, you can streamline the design process in MicroStation, improve model consistency, and maintain high efficiency in repetitive tasks. Cells, when used properly, become powerful tools in your CAD workflow, making large projects easier to manage and navigat

10. Manipulating Elements

Once elements like lines, circles, and arcs are created, they can be manipulated programmatically. You can move, rotate, or scale these elements using their respective properties and methods. This capability enables dynamic and responsive design processes within your MicroStation environment.

Moving an Element Using Coordinates in MicroStation VBA

IN MICROSTATION VBA, elements can be moved by calculating an offset between two points. The Move method requires an offset as a Point3d type, which represents the X, Y, and Z coordinates. This section explains how to move an element using a calculated offset.

Subroutine to Move an Element

THE FOLLOWING VBA CODE moves an element from one point to another by calculating the offset vector between the start and end points:

VBA Code 45

```
SUB ELEMENTMOVE()
    Dim oElement As Element
    Dim elementID As Long
    ' The element ID should be a numeric value, not a string
    Dim startPoint As Point3d
    Dim endPoint As Point3d
    Dim moveOffset As Point3d
    startPoint = Point3dFromXYZ(0, 0, 0)
    ' Starting point at the origin
    endPoint = Point3dFromXYZ(100, 100, 0)
    ' Destination point
    elementID = 467719
    ' Assign the actual ID of the element as a number
```

```vba
' Get the element by its ID
Set oElement = ActiveModelReference.GetElementByID64(elementID)
' Calculate the move offset by subtracting the
'startPoint coordinates from endPoint coordinates
moveOffset = Point3dFromXYZ(endPoint.X - startPoint.X, _
endPoint.Y - startPoint.Y, endPoint.Z - startPoint.Z)
' Move the element by the calculated move offset
oElement.Move moveOffset
' Rewrite the element in the model
oElement.Rewrite
End Sub
```

Key Concepts:

1. ELEMENT ID:

- The elementID is a unique identifier for an element in the model. You can retrieve it using MicroStation tools and use it to reference specific elements in your code.

2. Start and End Points:

- startPoint and endPoint are defined using Point3dFromXYZ. These represent the initial and final coordinates for the movement. In this case, the element is moved from (0, 0, 0) to (100, 100, 0).

3. Calculating the Offset:

- The offset is calculated by subtracting the startPoint coordinates from the endPoint coordinates. This results in the distance (or vector) by which the element will be moved. The moveOffset is a Point3d object, and it is passed to the Move method.

4. Using the Move Method:

- The Move method takes a Point3d offset and moves the element by this distance. It does not move the element to an absolute position but by a relative offset.

5. Rewriting the Element:

- After moving the element, the Rewrite method is used to update the element in the model so that the changes are reflected.

Practical Use Case:

THIS TECHNIQUE IS PARTICULARLY useful when needing to shift an element by a specific amount in MicroStation, for instance when performing adjustments based on user input or design changes.

By understanding how to draw and manipulate basic geometric shapes in VBA, you can leverage these foundational skills to create complex designs and automate repetitive tasks in MicroStation.

In addition to moving elements, MicroStation VBA provides powerful methods to **rotate** and **mirror** elements. This section explains how to perform these operations using VBA.

Rotating an Element

TO ROTATE AN ELEMENT, you can use the Rotate method along with a rotation matrix. The rotation matrix is created using the Matrix3dFromAxisAndRotationAngle function.

Subroutine to Rotate an Element:

VBA Code 46

```
SUB ROTATEELEMENT()
    Dim oElement As Element
    Dim elementID As Long
    Dim rotationPoint As Point3d
    Dim rotationAngleX As Double
    Dim rotationAngleY As Double
    Dim rotationAngleZ As Double
    elementID = 467728 ' Place your Element ID here
    Set oElement = ActiveModelReference.GetElementByID64(elementID)
    rotationPoint = Point3dFromXYZ(0, 0, 0)
    ' Rotation center point
```

```
' Define rotation angles for each axis (in radians)
rotationAngleX = 0 ' No rotation around the X axis
rotationAngleY = 0 ' No rotation around the Y axis
rotationAngleZ = 45 * Pi / 180
' 45 degrees rotation around the Z axis (converted to radians)
' Rotate the element using the pivot point and separate
'rotation angles for X, Y, and Z
oElement.Rotate rotationPoint, rotationAngleX, _
rotationAngleY, rotationAngleZ
' Rewrite the element in the model
oElement.Rewrite
End Sub
```

Key Concepts:

- **Rotation Matrix**: The Matrix3dFromAxisAndRotationAngle function creates a matrix for rotation around a specific axis (2 is for Z-axis).
- **Rotation Point**: The element is rotated around this point.
- **Angle in Radians**: The angle must be in radians, so degrees are converted using the formula degrees * Pi / 180.

Mirroring an Element

MIRRORING AN ELEMENT is done using the Mirror method. This requires two points that define the mirror line.

Subroutine to Mirror an Element:

VBA Code 47

```
SUB MIRRORELEMENT()
    Dim oElement As Element
    Dim elementID As Long
    Dim mirrorPoint1 As Point3d
    Dim mirrorPoint2 As Point3d
    elementID = 467719 ' Element ID
    Set oElement = ActiveModelReference.GetElementByID64(elementID)
```

```
mirrorPoint1 = Point3dFromXYZ(0, 0, 0)
' First point on the mirror line
mirrorPoint2 = Point3dFromXYZ(100, 0, 0)
' Second point on the mirror line
' Mirror the element
oElement.Mirror mirrorPoint1, mirrorPoint2
' Rewrite the element in the model
oElement.Rewrite
End Sub
```

Key Concepts:

- **Mirror Line**: Defined by two points, the element is mirrored across this line.
- **Mirror Operation**: This reflects the element as if it were placed across a mirror lying on the defined line.

Summary of Operations:

- **Rotation**: The element is rotated around a point by defining a rotation matrix.
- **Mirroring**: The element is mirrored across a line defined by two points.

Practical Applications:

THESE OPERATIONS ARE essential in various design and drafting scenarios. For example, rotation is often used in creating symmetrical designs, while mirroring can be helpful in duplicating elements across an axis.

11.
Grouping and Ungrouping Elements Using VBA Commands

In MicroStation, you can group multiple elements together, allowing them to be treated as a single entity, and later ungroup them to work with individual elements. Grouping is useful for working with assemblies, layouts, or repeated design components. This section covers how to programmatically group elements using their IDs and how to ungroup them using VBA.

1. Grouping Elements

TO GROUP ELEMENTS USING VBA, you can retrieve the elements by their IDs, select them in the model, and then issue the **GROUP** command. Grouping combines the selected elements so that they can be manipulated as a single object.

Subroutine to Group Elements by Their IDs:

VBA Code 48

```
SUB GROUPELEMENTS()
    Dim oElement1 As Element
    Dim oElement2 As Element
    Dim oElement3 As Element
    Dim oElement4 As Element
    Dim elementID1 As Long
    Dim elementID2 As Long
    Dim elementID3 As Long
    Dim elementID4 As Long
    ' Assign the IDs of the elements
    elementID1 = 467868
    elementID2 = 467869
    elementID3 = 467870
    elementID4 = 467871
    ' Get the elements by their IDs
```

```
Set oElement1 = _
ActiveModelReference.GetElementByID64(elementID1)
Set oElement2 = _
ActiveModelReference.GetElementByID64(elementID2)
Set oElement3 = _
ActiveModelReference.GetElementByID64(elementID3)
Set oElement4 = _
ActiveModelReference.GetElementByID64(elementID4)
' Select the elements to group
ActiveModelReference.SelectElement oElement1, True
ActiveModelReference.SelectElement oElement2, True
ActiveModelReference.SelectElement oElement3, True
ActiveModelReference.SelectElement oElement4, True
' Group the selected elements using the GROUP command
CadInputQueue.SendCommand "GROUP"
' Unselect all elements after grouping
ActiveModelReference.UnselectAllElements
CommandState.StartDefaultCommand
End Sub
```

Key Concepts:

- **Element Selection**: The elements are retrieved using their IDs with ActiveModelReference.GetElementByID64, and then they are selected using SelectElement.
- **GROUP Command**: The CadInputQueue.SendCommand "GROUP" command is used to group the selected elements. This makes the elements behave as a single object.
- **Unselecting Elements**: After grouping, the elements are deselected using UnselectAllElements to clean up the selection in the user interface.

2. Ungrouping Elements

UNGROUPING ELEMENTS that were previously grouped can be done using the **UNGROUP** command. This breaks the group back into individual components so that each element can be manipulated separately.

Subroutine to Ungroup Elements:

VBA Code 49

```
SUB UNGROUPCELL()
    Dim oCell As Element
    Dim elementID As Long
    elementID = 467872
    ' Assign the ID of the cell or grouped element
    ' Get the grouped element by its ID
    Set oCell = ActiveModelReference.GetElementByID64(elementID)
    ' Select the element to ungroup
    ActiveModelReference.SelectElement oCell, True
    ' Ungroup the selected element using the UNGROUP command
    CadInputQueue.SendCommand "UNGROUP"
    ' Unselect all elements after ungrouping
    ActiveModelReference.UnselectAllElements
    CommandState.StartDefaultCommand
End Sub
```

Key Concepts:

- **Selecting the Grouped Element**: The group or cell is retrieved by its ID and selected using SelectElement.
- **UNGROUP Command**: The CadInputQueue.SendCommand "UNGROUP" command is used to break apart the grouped elements, allowing them to be edited individually again.
- **Cleaning Up**: After ungrouping, the elements are deselected using UnselectAllElements to return the user interface to its default state.

3. Combining Grouping and Ungrouping

GROUPING AND UNGROUPING elements are complementary operations that can be applied dynamically based on design needs. Grouping allows you to temporarily combine multiple elements into a single entity for easier manipulation, while ungrouping restores individual control over each element.

Summary of Grouping and Ungrouping Operations:

- **Grouping**: Use SelectElement to select multiple elements, and then group them with the GROUP command. This allows the selected elements to be treated as one.
- **Ungrouping**: Use the UNGROUP command to break grouped elements back into their individual components. This restores the ability to manipulate each element separately.

Practical Applications:

- **Grouping**: This is useful for combining elements such as furniture in architectural drawings, assemblies in mechanical designs, or other collections of elements that need to be moved or manipulated as one.
- **Ungrouping**: When you need to modify individual elements within a group, such as updating the position of a component within an assembly, you can ungroup the elements, make the necessary changes, and regroup them if needed.

12. Retrieving Element IDs in MicroStation VBA

In MicroStation VBA, you can retrieve an element's unique identifier (ID) and perform operations such as scanning through all elements or retrieving specific elements using their IDs. The ID64 property is used to access the 64-bit unique identifier of an element. This section explains how to retrieve element IDs using both a scanning approach and by directly accessing an element by its ID.

1. Retrieving Element IDs by Scanning the Model

YOU CAN LOOP THROUGH all the elements in the active model and retrieve their IDs using the ID64 property. Here's how you can scan the model and print the IDs of all visible elements.

Subroutine to Retrieve Element IDs via Scan:

VBA Code 50

```
SUB RETRIEVEELEMENTID()
    Dim oElement As Element
    Dim oEnum As ElementEnumerator
    Dim elementID As Variant
    ' Use Variant to handle potential large values from ID64
    ' Set scan criteria if needed (optional)
    Dim MyScanCriteria As ElementScanCriteria
    Set MyScanCriteria = New ElementScanCriteria
    MyScanCriteria.IncludeOnlyVisible
    ' Optional: include only visible elements
    ' Scan the active model
    Set oEnum = ActiveModelReference.Scan(MyScanCriteria)
    ' Loop through the elements found by the scan
    Do While oEnum.MoveNext
        Set oElement = oEnum.Current
        ' Check if the element is valid
        If Not oElement Is Nothing Then
```

```
' Safely retrieve the element ID64
On Error Resume Next
elementID = oElement.ID64
' Retrieve the 64-bit element ID
On Error GoTo 0
' Print the ID in the Immediate Window if no error
If elementID <> Empty Then
Debug.Print "Element ID64: " & elementID
Else
Debug.Print _
"Could not retrieve ID64 for this element."
End If
End If
Loop
End Sub
```

Key Concepts:

- **Element Enumerator (oEnum)**: The enumerator allows you to loop through the elements in the active model.
- **ID64**: The ID64 property retrieves the 64-bit unique identifier of each element.
- **Scan Criteria**: You can use ElementScanCriteria to filter the elements you want to scan, such as including only visible elements.

2. Retrieving Specific Elements by ID

IF YOU KNOW THE ID64 of an element, you can retrieve it directly using the GetElementByID64 method. This is useful when you need to manipulate a specific element by its unique ID.

Subroutine to Retrieve Element by Its ID:

VBA Code 51

```
SUB RETRIEVEELEMENTBYID()
    Dim oElement As Element
    Dim elementID As Long
```

```
elementID = 467868
' Replace with the actual ID of the element
' Get the element by its ID
Set oElement = ActiveModelReference.GetElementByID64(elementID)
' Check if the element exists
If Not oElement Is Nothing Then
Debug.Print "Element with ID " & elementID & " exists."
Else
MsgBox "Element with ID " & elementID & " not found."
End If
End Sub
```

Key Concepts:

- **GetElementByID64**: This method retrieves an element using its unique ID64. It returns Nothing if the element is not found.
- **Error Handling**: You can include error handling or conditional checks to ensure the element is valid before proceeding.

Best Practices for Retrieving Element IDs

1. **Use ID64 for Large Models**: The ID64 property is essential for retrieving unique identifiers in large models that require 64-bit IDs.
2. **Error Handling**: Always ensure that the element exists before attempting to retrieve its ID or perform operations on it. Use error handling to manage cases where an element cannot be found.
3. **Scan Criteria**: When scanning through elements, it's useful to apply filtering using ElementScanCriteria to reduce the number of elements processed, such as by only scanning visible or unlocked elements.

Practical Applications:

- **Scanning for Specific Elements**: Use the scan method when you want to loop through all elements in a model and work with their properties.
- **Direct Element Retrieval**: If you need to manipulate a specific element by its ID, use GetElementByID64 to retrieve it directly and

ensure it exists before proceeding.

THESE TECHNIQUES PROVIDE a flexible way to retrieve and manipulate elements in MicroStation VBA, allowing you to work with both large groups of elements and individual elements by their unique IDs.

13.
Calling Subroutines in VBA

Subroutines, also known as procedures, are a core part of VBA programming. They allow you to write reusable code blocks that can be called from other parts of the program. This chapter explains how to define, call, and pass arguments to subroutines, how to use the Call keyword to invoke subroutines, and best practices for organizing your VBA code.

Understanding Subroutines and Arguments

A **subroutine** is a block of code that performs a specific task. It can be reused throughout your VBA code by calling it whenever needed. Subroutines can also take **arguments**, which allow you to pass data into the subroutine and customize its behaviour based on the inputs.

Example: Calling a Subroutine with Arguments

VBA Code 52

```
SUB MAIN_RECTANGLE()
    ' Calling the DrawRectangle subroutine with arguments
    DrawRectangle 10, 20
    End Sub
    Sub DrawRectangle(width As Double, height As Double)
    ' Code to draw a rectangle based on the width and height passed
    Area = height * width
    Debug.Print "Drawing a rectangle with width: " & width _
    & " and height: " & height
    Debug.Print "Area is " & Area
    End Sub
```

In this example:

- **Subroutine Declaration**: Sub DrawRectangle(width As Double, height As Double) declares a subroutine that accepts two arguments (width and height).

- **Calling the Subroutine**: In the Main subroutine, DrawRectangle 10, 20 calls the DrawRectangle subroutine with specific values (10 and 20) passed as arguments.
- **Result**: The subroutine performs the task of drawing a rectangle (or, in this case, printing the dimensions).

Benefits of Using Subroutines with Arguments:

- **Code Reusability**: By passing different values to the same subroutine, you can reuse the same code for different scenarios.
- **Modularity**: Subroutines help break your code into smaller, manageable parts that are easier to maintain and debug.

Calling Subroutines Using the Call Keyword

IN VBA, YOU CAN USE the Call keyword to explicitly invoke a subroutine. Although not strictly necessary, using Call can make your code more readable, especially when working with subroutines that take multiple arguments. If the subroutine accepts arguments, they must be enclosed in parentheses when using Call.

Example: Using the Call Keyword to Invoke a Subroutine

VBA Code 53

```
SUB MAIN_CIRCLE()
    ' Calling the subroutine using Call with
    'parentheses around arguments
    Call DrawCircle(5)
    End Sub
    Sub DrawCircle(radius As Double)
    ' Code to draw a circle based on the radius passed
    Debug.Print "Drawing a circle with radius: " & radius
    End Sub
```

In this example:

- **Call Keyword**: The Call keyword is used to invoke the DrawCircle subroutine, passing the value 5 as the radius.
- **Parentheses**: When using the Call keyword, parentheses are required around the arguments.

When to Use the Call Keyword:

- **Readability**: It can make the subroutine call more explicit and easier to spot in complex code.
- **Compatibility**: When refactoring code or converting from other languages that use Call, it can be useful to retain the explicit structure.

Calling Subroutines from Other Code Blocks

VBA ALLOWS YOU TO CALL subroutines from within other subroutines or functions. This technique helps to organize complex tasks by breaking them into smaller, more manageable operations. You can also **chain subroutine calls**, where one subroutine calls another, allowing for nested operations and efficient code reuse.

Example: Chaining Subroutine Calls

VBA Code 54

```
SUB MAINPROCEDURE()
    ' Calling other subroutines within MainProcedure
    Call CreateLevel("DesignLevel")
    ' Using Call to invoke subroutine
    DrawShape
End Sub
Sub CreateLevel(levelName As String)
    ' Code to create a level
    Debug.Print "Creating level: " & levelName
End Sub
Sub DrawShape()
    ' Code to draw a shape
    Debug.Print "Drawing a shape"
End Sub
```

In this example:

- **Chaining Subroutine Calls**: The MainProcedure subroutine calls CreateLevel (using Call) and DrawShape.
- **Using Call**: The CreateLevel subroutine is explicitly called using Call, while DrawShape is called without Call. Both methods work, but Call makes the code more explicit.

Using Subroutines for Code Modularity

SUBROUTINES ARE FUNDAMENTAL to achieving **modular programming** in VBA. Modular code is easier to read, debug, and maintain because each subroutine handles a specific task. This approach allows you to:

- **Test** individual subroutines in isolation.
- **Reuse** subroutines across different parts of your project.
- **Extend** functionality by adding more subroutines without affecting existing ones.

Example: Modular Code with Multiple Subroutines

VBA Code 55

```
SUB MAINTASK()
    InitializeSettings
    ProcessData
    FinalizeTask
End Sub
Sub InitializeSettings()
    ' Code to initialize settings
    Debug.Print "Initializing settings..."
End Sub
Sub ProcessData()
    ' Code to process data
    Debug.Print "Processing data..."
End Sub
Sub FinalizeTask()
    ' Code to finalize the task
    Debug.Print "Finalizing task..."
End Sub
```

In this example:

- **Modular Design**: Each subroutine (InitializeSettings, ProcessData, and FinalizeTask) is responsible for a specific part of the task, making the code easier to maintain.
- **Main Task**: The MainTask subroutine coordinates the entire process by calling these subroutines in sequence.

Best Practices for Calling Subroutines

1. **Use Descriptive Names**: Always use clear, descriptive names for your subroutines so that their purpose is immediately clear. For example, use DrawRectangle instead of Draw or Rectangle.
2. **Limit the Number of Arguments**: While passing arguments to subroutines is helpful, avoid passing too many arguments, as this can make your subroutine calls complicated. If needed, group related

arguments into objects or use optional parameters.
3. **Document Your Subroutines**: Add comments to explain the purpose of each subroutine, especially when passing arguments. This helps others (or yourself) understand the logic later.
4. **Avoid Side Effects**: Subroutines should perform their task without unintended side effects (e.g., modifying global variables unexpectedly). Keeping subroutines focused on a specific task makes debugging easier.
5. **Use Exit Sub to Handle Errors**: If a condition is met that requires the subroutine to end early, use Exit Sub to exit the subroutine cleanly.

Summary of Subroutine Usage in VBA

- **Subroutines** are reusable blocks of code that perform specific tasks. They can take arguments to customize their behaviour.
- **Chaining subroutines** by calling one subroutine from another allows you to break down complex tasks into smaller, more manageable parts.
- **Modular programming** with subroutines makes your code easier to read, debug, and extend.
- **Call Keyword**: You can use the Call keyword to explicitly invoke subroutines, especially when passing arguments.

BY USING SUBROUTINES effectively, you can write cleaner, more maintainable VBA code, enhancing the overall structure and performance of your programs.

Exercise 2: Working with Multiple Elements

IN THIS EXERCISE, YOU will create and manipulate multiple elements (such as lines, circles, and rectangles) in MicroStation using VBA. You will also learn how to group elements and perform operations on groups of elements. The focus will be on using loops, subroutines, and efficient techniques to manage multiple elements in the active model.

1. Creating and Managing Multiple Elements

WHEN WORKING WITH MULTIPLE elements, loops are often used to create and manipulate several elements in sequence. For instance, you can create multiple lines or circles by iterating through a loop and specifying different start and end points for each element.

Example: Drawing Multiple Lines

VBA Code 56

```
SUB DRAWMULTIPLELINES()
    Dim oLine As LineElement
    Dim startPoint As Point3d
    Dim endPoint As Point3d
    Dim i As Integer
    ' Loop to create 5 lines with increasing X-coordinates
    For i = 1 To 5
    ' Define start and end points for each line
    startPoint = Point3dFromXYZ(i * 10, 0, 0)
    endPoint = Point3dFromXYZ(i * 10, 50, 0)
    ' Create a line element
    Set oLine = CreateLineElement2 _
    (Nothing, startPoint, endPoint)
    ' Add the line to the active model
    ActiveModelReference.AddElement oLine
    Next i
End Sub
```

In this example:

- **For Loop**: The loop runs five times, creating five separate lines.
- **Dynamic Start/End Points**: The X-coordinate of the start and end points changes with each iteration, drawing parallel lines.
- **Efficiency**: Instead of manually specifying each line, the loop automates the process.

Practical Tip:

- Use loops to efficiently create and place multiple elements in the model, especially when dealing with repetitive structures such as grids or arrays of shapes.

2. Grouping and Manipulating Elements

AFTER CREATING MULTIPLE elements, you may want to group them for easier manipulation. MicroStation VBA allows you to group elements together using the ComplexElement or CellElement objects. Once grouped, elements can be moved, rotated, or deleted as a single unit.

Example: Grouping Multiple Lines into a Cell

VBA Code 57

```
SUB GROUPMULTIPLELINESASCELL()
    Dim oLine As LineElement
    Dim oCell As CellElement
    Dim elements(0 To 4) As Element
    Dim startPoint As Point3d
    Dim endPoint As Point3d
    Dim i As Integer
    ' Create 5 lines and store them in an array
    For i = 0 To 4
    startPoint = Point3dFromXYZ(i * 10, 0, 0)
    endPoint = Point3dFromXYZ(i * 10, 50, 0)
    ' Create the line element
    Set oLine = CreateLineElement2_
    (Nothing, startPoint, endPoint)
    ' Cast the LineElement as an Element before
    'storing it in the array
    Set elements(i) = oLine
    Next i
    ' Group the lines into a cell element
    Set oCell = CreateCellElement1("LineGroup", _
    elements, Point3dFromXYZ(0, 0, 0))
    ' Add the cell (grouped lines) to the active model
```

```
ActiveModelReference.AddElement oCell
End Sub
```

In this example:

- **Element Array**: Each line element is stored in an array, allowing for easy grouping.
- **CreateCellElement1**: This method is used to group the elements into a single cell, which can be manipulated as a whole.
- **Reusability**: Once grouped, the cell can be reused or manipulated in other operations.

Practical Tip:

- Group related elements into cells to simplify manipulation (such as moving, copying, or rotating them as a single entity).

3. Manipulating Groups of Elements

ONCE YOU HAVE GROUPED elements together, you can manipulate the group by moving, rotating, or scaling it. This is especially useful when dealing with assemblies or collections of related elements.

Example: Moving a Group of Elements

VBA Code 58

```
SUB MOVEGROUPEDELEMENTS()
    Dim oElement As Element
    Dim moveOffset As Point3d
    ' Define the movement vector (e.g., move 20 units in the X direction)
    moveOffset = Point3dFromXYZ(-20, 0, 0)
    ' Select all elements in the model
    ' (for this example, we assume all are grouped)
    Set oElement = ActiveModelReference.GetElementByID64(468021)
    ' Replace with actual ID of your group
    ' Move the selected group by the defined vector
```

```
If Not oElement Is Nothing Then
oElement.Move moveOffset
oElement.Rewrite ' Commit the change to the model
End If
End Sub
```

In this example:

- **Movement Offset**: The group is moved by 20 units in the X direction.
- **Move Method**: The Move method applies the movement offset to the entire group of elements.
- **Rewriting the Element**: After moving the group, Rewrite is called to commit the changes to the active model.

Practical Tip:

- Use the Move, Rotate, and Scale methods to manipulate groups of elements efficiently. Always call Rewrite to save changes to the model.

Best Practices for Working with Multiple Elements

1. **Efficient Loops**: When creating multiple elements, ensure your loops are efficient and avoid unnecessary operations within the loop.
2. **Element Grouping**: Group elements into cells or complex elements for easier management, especially when dealing with assemblies or repeated components.
3. **Element Manipulation**: Use vectors and matrices to move, rotate, and scale elements programmatically. This is faster and more accurate than manually adjusting elements.
4. **ElementEnumerator**: When working with large groups of elements, use ElementEnumerator to efficiently iterate over and manipulate collections of elements.

Exercise: Drawing and Grouping Elements

OBJECTIVE:

- Write a VBA routine that creates multiple elements (lines, circles, or rectangles), groups them, and manipulates the group as a whole.

Steps:

1. **Create Multiple Elements**: Write a subroutine that draws 3 circles and 3 lines in a grid pattern.
2. **Group the Elements**: Group the created circles and lines into a single cell.
3. **Move the Group**: Write another subroutine to move the entire group of elements 50 units in the X direction.

Hints:

- Use loops to create the elements in a grid pattern.
- Store the elements in an array for easy grouping.
- Use the Move method to translate the group.

Solution Example:

VBA Code 59

```
SUB MAINPROCEDUREGROUPANDMOVE()
    Dim ElementID As Double
    ' Calling other subroutines within MainProcedure
    DrawAndGroupElements ElementID
    ' Pass the ID of the group created
    MoveShapeGroup ElementID
    ' Move the group using the retrieved ID
    End Sub
    Sub DrawAndGroupElements(ByRef ElementID As Double)
    Dim oCircle As EllipseElement
    Dim oLine As LineElement
    Dim oCell As CellElement
```

```
Dim elements(0 To 5) As Element
Dim i As Integer
' Create 3 circles
For i = 0 To 2
Set oCircle = _
CreateEllipseElement2 _
(Nothing, Point3dFromXYZ(i * 10, i * 10, 0), 5, 5, _
Matrix3dIdentity)
Set elements(i) = oCircle
Next i
' Create 3 lines
For i = 3 To 5
Set oLine = CreateLineElement2 _
(Nothing, Point3dFromXYZ _
(i * 10 - 30, 0, 0), Point3dFromXYZ(i * 10 - 30, 50, 0))
Set elements(i) = oLine
Next i
' Group the elements into a cell
Set oCell = CreateCellElement1 _
("ShapeGroup", elements, Point3dFromXYZ(0, 0, 0))
ActiveModelReference.AddElement oCell
' Retrieve the ID of the cell element
'and pass it back via the ByRef parameter
ElementID = oCell.ID64
End Sub
Sub MoveShapeGroup(ElementID As Double)
Dim oElement As Element
Dim moveOffset As Point3d
' Define the movement vector
' (e.g., move 20 units in the X direction)
moveOffset = Point3dFromXYZ(20, 0, 0)
' Get the grouped element by its ID
Set oElement = ActiveModelReference.GetElementByID64(ElementID)
' Move the selected group by the defined vector
If Not oElement Is Nothing Then
oElement.Move moveOffset
oElement.Rewrite ' Commit the change to the model
End If
End Sub
```

Solution Explanation

WE WILL BREAK DOWN the process into three subroutines:

1. **Creating and Grouping the Elements**: Drawing the elements and grouping them as a cell.
2. **Retrieving the Group's Element ID**: Using the ID64 property to identify the grouped elements.
3. **Moving the Grouped Elements**: Moving the entire group of elements based on their ID64.

1. Main Procedure: Combining Creation and Movement

THIS IS THE MAIN SUBROUTINE that ties everything together. It first creates and groups the elements by calling DrawAndGroupElements and then moves the group using MoveShapeGroup.

VBA Code 60

```
SUB MAINPROCEDUREGROUPANDMOVE()
    Dim ElementID As Double
    ' Call subroutine to draw and group elements,
    'retrieving the group ElementID
    DrawAndGroupElements ElementID
    ' Call subroutine to move the group using
    'the retrieved ElementID
    MoveShapeGroup ElementID
    End Sub
```

Explanation:

- **MainProcedureGroupandMove** is the control procedure that first creates and groups the elements, then moves the grouped elements by passing the ElementID between subroutines.
- The **ElementID** is a Double because ID64 is a large number, and it is passed by reference to ensure the correct element is identified.

2. Drawing and Grouping Elements into a Cell

THIS SUBROUTINE CREATES three circles and three lines, groups them into a cell, and retrieves the ID64 of the grouped cell.

VBA Code 61

```
SUB DRAWANDGROUPELEMENTS(ByRef ElementID As Double)
    Dim oCircle As EllipseElement
    Dim oLine As LineElement
    Dim oCell As CellElement
    Dim elements(0 To 5) As Element
    Dim i As Integer
    ' Create 3 circles
    For i = 0 To 2
    Set oCircle = _
    CreateEllipseElement2 _
    (Nothing, Point3dFromXYZ(i * 10, i * 10, 0), 5, 5, _
    Matrix3dIdentity)
    Set elements(i) = oCircle
    Next i
    ' Create 3 lines
    For i = 3 To 5
    Set oLine = CreateLineElement2 _
    (Nothing, Point3dFromXYZ _
    (i * 10 - 30, 0, 0), Point3dFromXYZ(i * 10 - 30, 50, 0))
    Set elements(i) = oLine
    Next i
    ' Group the elements into a cell
    Set oCell = CreateCellElement1 _
    ("ShapeGroup", elements, Point3dFromXYZ(0, 0, 0))
    ActiveModelReference.AddElement oCell
    ' Retrieve the ID of the cell element and pass it
    'back via the ByRef parameter
    ElementID = oCell.ID64
    End Sub
```

Explanation:

- **Element Creation**: The subroutine creates three circles and three

lines using a loop. Each element is stored in an array of Element type (elements(0 To 5)).
- **Grouping Elements**: After the elements are created, they are grouped into a single cell using the CreateCellElement1 method. The array of elements is passed as an argument to group them together.
- **Retrieving the ElementID**: Once the cell is created, its ID64 is retrieved. This is the unique identifier of the group, which is passed back to the main subroutine via the ByRef parameter.

3. Moving the Grouped Elements

THIS SUBROUTINE MOVES the group of elements by referencing their ElementID. The group is moved 20 units in the X direction.

VBA Code 62

```
SUB MOVESHAPEGROUP(ElementID As Double)
    Dim oElement As Element
    Dim moveOffset As Point3d
    ' Define the movement vector
    ' (e.g., move 20 units in the X direction)
    moveOffset = Point3dFromXYZ(-20, 0, 0)
    ' Get the grouped element by its ID
    Set oElement = ActiveModelReference.GetElementByID64(ElementID)
    ' Move the selected group by the defined vector
    If Not oElement Is Nothing Then
    oElement.Move moveOffset
    oElement.Rewrite ' Commit the change to the model
    End If
End Sub
```

Explanation:

- **Movement Vector**: The group is moved by 20 units in the X direction using the Point3dFromXYZ function to define a movement vector (moveOffset).
- **Retrieving the Group by ID**: The GetElementByID64 method

retrieves the group (cell) by its ElementID. This ensures that the correct group of elements is selected for manipulation.
- **Moving and Saving Changes**: Once the group is retrieved, the Move method moves it by the specified vector. The Rewrite method commits the changes to the active model.

Summary of Key Concepts:

- **Modular Subroutines**: Each task—drawing, grouping, and moving elements—is broken down into individual subroutines to improve code clarity and reusability.
- **Passing ElementID by Reference**: The ElementID of the grouped cell is passed between subroutines using the ByRef keyword. This ensures that the correct group is identified and moved.
- **Grouping Elements into a Cell**: Grouping elements into a cell allows for easier manipulation of multiple elements as a single unit. This is especially useful when working with complex assemblies or repeated elements in a model.
- **Efficient Element Manipulation**: By retrieving and moving elements based on their ID64, you can efficiently manipulate specific groups without needing to scan the entire model repeatedly.

Practical Application:

THIS EXERCISE DEMONSTRATES how to manage multiple elements efficiently by creating, grouping, and manipulating them as a whole. These techniques are crucial in large-scale designs where individual manipulation of each element would be cumbersome and error-prone.

By using grouping and element IDs, you can simplify your workflow and automate the manipulation of complex structures within Micro Station VBA.

14. Element Locking and Unlocking

In MicroStation, elements can be locked programmatically to prevent unintended modifications, deletions, or movements. Locking is a useful feature when working in collaborative environments or when you need to protect specific elements in a design. This section covers how to lock and unlock elements using VBA, along with best practices for managing locked elements in your designs.

1. Locking Elements Programmatically

LOCKING AN ELEMENT ensures that it cannot be moved, modified, or deleted until it is explicitly unlocked. The IsLocked property of the element determines its lock state. By setting this property to True, you can lock the element.

Example: Locking an Element

VBA Code 63

```
SUB LOCKELEMENT()
    Dim oElement As Element
    ' Retrieve the element by its ID
    Set oElement = ActiveModelReference.GetElementByID64(123456789) ' Replace with actual Element ID
    ' Lock the element
    oElement.IsLocked = True
    ' Write the changes back to the model
    oElement.Rewrite
End Sub
```

Explanation:

- **GetElementByID64**: This function retrieves an element by its unique ID64 (a 64-bit identifier).

- **IsLocked Property**: Setting IsLocked = True locks the element, preventing further modification or deletion.
- **Rewrite**: After making changes to the element's properties (in this case, locking it), you must call Rewrite to commit the changes back to the active model.

2. Unlocking Elements via VBA

TO UNLOCK AN ELEMENT, you simply set the IsLocked property to False. This allows the element to be modified or deleted once again.

Example: Unlocking an Element

VBA Code 64

```
SUB UNLOCKELEMENT()
    Dim oElement As Element
    ' Retrieve the element by its ID
    Set oElement = ActiveModelReference.GetElementByID64(123456789) ' Replace with actual Element ID
    ' Unlock the element
    oElement.IsLocked = False
    ' Write the changes back to the model
    oElement.Rewrite
    End Sub
```

Explanation:

- **Unlocking the Element**: By setting IsLocked = False, the element is unlocked, allowing further modifications.
- **Rewrite Changes**: The Rewrite method ensures that the changes (in this case, unlocking the element) are written back to the model so that the element's new state persists.

3. Locking and Unlocking Multiple Elements

IN PRACTICAL USE CASES, you may need to lock or unlock multiple elements in a design. You can achieve this by using the ElementEnumerator to iterate through a set of elements and apply the lock or unlock operation to each.

Example: Locking Multiple Elements

VBA Code 65

```
SUB LOCKALLELEMENTS()
    Dim oElement As Element
    Dim oEnum As ElementEnumerator
    Dim MyScanCriteria As ElementScanCriteria
    ' Set up scan criteria to exclude non-graphical elements
    ' (optional),
    Set MyScanCriteria = New ElementScanCriteria
    MyScanCriteria.ExcludeNonGraphical
    ' Scan the active model and retrieve all elements
    Set oEnum = ActiveModelReference.Scan(MyScanCriteria)
    ' Loop through all elements and lock each one
    Do While oEnum.MoveNext
    Set oElement = oEnum.Current
    ' Check if the element can be locked
    On Error Resume Next ' Handle potential errors gracefully
    If Not oElement.IsLocked Then
    oElement.IsLocked = True
    oElement.Rewrite ' Commit the change to the model
    End If
    On Error GoTo 0 ' Restore default error handling
    Loop
    End Sub
```

Example: Unlocking Multiple Elements

VBA Code 66

```
SUB UNLOCKALLELEMENTS()
    Dim oElement As Element
```

```
Dim oEnum As ElementEnumerator
Dim MyScanCriteria As ElementScanCriteria
' Set up scan criteria to exclude non-graphical elements
' (optional)
Set MyScanCriteria = New ElementScanCriteria
MyScanCriteria.ExcludeNonGraphical
' Scan the active model and retrieve all elements
Set oEnum = ActiveModelReference.Scan(MyScanCriteria)
' Loop through all elements and lock each one
Do While oEnum.MoveNext
Set oElement = oEnum.Current
' Check if the element can be unlocked
On Error Resume Next ' Handle potential errors gracefully
If oElement.IsLocked Then
oElement.IsLocked = False
oElement.Rewrite ' Commit the change to the model
End If
On Error GoTo 0 ' Restore default error handling
Loop
End Sub
```

Explanation:

- **ElementEnumerator**: The ElementEnumerator object allows you to iterate over all elements in the model. For each element, you can set the IsLocked property to True (for locking) or False (for unlocking).
- **Efficient Locking/Unlocking**: This method allows you to lock or unlock a large number of elements efficiently, especially in complex designs where individual element manipulation would be cumbersome.

4. Best Practices for Locking and Unlocking Elements

WHILE LOCKING ELEMENTS can be helpful in protecting important components of your design, it is important to use this feature judiciously to avoid potential complications.

Key Best Practices:

1. **Lock Only When Necessary**: Only lock elements that you are certain should not be modified. Excessive locking of elements can make the design harder to manipulate later.
2. **Track Locked Elements**: If you plan to lock multiple elements, it's a good idea to document which elements are locked or implement a system to track them. This can prevent confusion in collaborative environments or large projects.
3. **Unlock When No Longer Needed**: Always remember to unlock elements once their protected state is no longer required. Failing to do so can cause problems for other users who may need to modify or delete those elements.
4. **Use IDs for Specific Element Control**: When working with specific elements (especially in collaborative settings), ensure you retrieve and lock the correct element by using its ID64. This prevents accidental locking or unlocking of unrelated elements.
5. **Efficiency in Bulk Operations**: When locking or unlocking multiple elements, ensure you perform these operations efficiently using loops and ElementEnumerator. This helps in managing large models without performance issues.

5. Handling Locked Elements in Collaborative Environments

WHEN WORKING IN ENVIRONMENTS where multiple users are interacting with the same design, locking elements can be useful to prevent unintended changes by others. However, this should be done with caution:

- **Notify Other Users**: Communicate with team members when elements are locked to avoid confusion.
- **Revisit Locked Elements**: Regularly review locked elements to ensure they still need to be in a locked state.

Summary of Key Concepts

- **Locking Elements**: Use the IsLocked = True property to lock elements and prevent them from being modified or deleted.
- **Unlocking Elements**: Unlock elements by setting IsLocked = False and writing the changes back to the model.
- **Best Practices**: Lock elements only when necessary, document locked elements, and unlock them when no longer needed.
- **Bulk Operations**: Use loops and ElementEnumerator to efficiently lock or unlock multiple elements in large models.

BY MASTERING THE LOCKING and unlocking of elements, you can protect key parts of your design from accidental changes while maintaining the flexibility to modify them when needed. These techniques are especially important in large-scale projects and collaborative environments where multiple users interact with the same design files.

15. Conditional Logic in VBA

Conditional logic allows your program to make decisions based on certain conditions. It is a fundamental concept in programming, and in VBA, conditional logic is primarily implemented using If...Then, ElseIf, Else, and Select Case statements. These structures let you control the flow of your program by executing different blocks of code depending on certain conditions.

1. Using If Statements

THE IF...THEN STATEMENT is one of the most common ways to implement conditional logic in VBA. It allows you to test whether a condition is True and execute a block of code if the condition is met.

Example: Basic If Statement

VBA Code 67

```
SUB CHECKVALUE()
    Dim value As Integer
    value = 10
    ' If the value is greater than 5, show a message box
    If value > 5 Then
    MsgBox "Value is greater than 5"
    End If
End Sub
```

Explanation:

- **Condition**: The code checks whether value > 5. If the condition is True, the message box is displayed.
- **Single Condition**: In this case, only one condition is being tested. If the condition is not met (i.e., if value is less than or equal to 5), the

code inside the If block will not be executed.

2. Using ElseIf and Else for Multiple Conditions

OFTEN, YOU WILL NEED to test multiple conditions. This can be done using **ElseIf** and **Else** clauses. These extensions of the If statement allow for more complex decision-making.

Example: Using ElseIf and Else

VBA Code 68

```
SUB EVALUATEVALUE()
    Dim value As Integer
    value = 10
    ' Multiple conditions to evaluate the value
    If value > 10 Then
    MsgBox "Value is greater than 10"
    ElseIf value = 10 Then
    MsgBox "Value is exactly 10"
    Else
    MsgBox "Value is less than 10"
    End If
End Sub
```

Explanation:

- **ElseIf**: Adds additional conditions to check. If value > 10 is False, VBA checks whether value = 10.
- **Else**: The Else clause acts as a catch-all. If none of the previous conditions are met, the code in the Else block is executed.
- **Decision Tree**: Only one block of code will be executed, depending on which condition is True.

3. Other Conditional Constructs: Select Case

FOR SCENARIOS WHERE you have many possible values to evaluate, the Select Case statement offers a cleaner alternative to multiple If...ElseIf statements. It is particularly useful when you need to compare a single variable against many different values.

Example: Using Select Case

VBA Code 69

```
SUB CHOOSECASE()
    Dim value As Integer
    value = 2
    ' Select case to evaluate multiple values
    Select Case value
    Case 1
    MsgBox "Value is 1"
    Case 2
    MsgBox "Value is 2"
    Case 3
    MsgBox "Value is 3"
    Case Else
    MsgBox "Value is something else"
    End Select
End Sub
```

Explanation:

- **Select Case**: The Select Case structure evaluates the value of a variable (value in this case) and compares it against several possible cases (1, 2, 3, etc.).
- **Case Else**: Similar to Else in an If statement, Case Else catches any cases that do not match the specified values.

When to Use Select Case:

- **Simpler Syntax**: Use Select Case when you have a single variable with

multiple possible values to evaluate. It makes the code more readable and avoids deeply nested If statements.
- **Complexity**: For simple conditions, If...Then works well. For a large number of conditions, Select Case is cleaner and easier to manage.

4. Best Practices for Conditional Logic

1. **Avoid Deep Nesting**: Deeply nested If statements make the code hard to read and maintain. Instead, break up the logic or use Select Case when appropriate.
2. **Short-Circuiting**: If you have multiple conditions to check, order them from most likely to least likely. This way, VBA can skip unnecessary checks, improving performance.
3. **Default Cases**: Always include a catch-all condition, such as Else or Case Else, to handle unexpected or default cases.
4. **Maintainability**: When your logic grows complex, consider breaking it into separate subroutines or functions. This helps with testing, debugging, and readability.

5. Combining Conditions

YOU CAN COMBINE MULTIPLE conditions in a single If statement using logical operators such as And, Or, and Not. This allows you to evaluate more than one condition at a time.

Example: Using Logical Operators

VBA Code 70

```
SUB COMBINECONDITIONS()
    Dim value1 As Integer
    Dim value2 As Integer
    value1 = 10
    value2 = 20
    ' Check if both conditions are true
    If value1 > 5 And value2 > 15 Then
```

```
MsgBox "Both values are greater than their thresholds"
End If
End Sub
```

Explanation:

- **And Operator**: Both conditions (value1 > 5 and value2 > 15) must be True for the code inside the If block to execute.
- **Other Logical Operators**:
    - Or: Only one of the conditions needs to be True.
    - Not: Negates the condition (i.e., it checks whether a condition is False).

EXERCISE: IMPLEMENTING Conditional Logic

In this exercise, you'll write a subroutine that evaluates a score and assigns a grade based on specific conditions.

Objective:

- Write a subroutine that assigns a grade (A, B, C, D, or F) based on a given score.

Steps:

1. **Create a Variable**: Define a variable score to hold the student's score.
2. **Use Conditional Logic**: Use If...ElseIf...Else or Select Case to assign the grade based on the score.
3. **Display the Grade**: Output the grade using MsgBox.

Solution Example:

VBA Code 71

```
SUB ASSIGNGRADE()
    Dim score As Integer
    Dim grade As String
    score = 85 ' You can change this value to test different scores
    ' Using If...ElseIf...Else to assign a grade based on the score
```

```
If score >= 90 Then
    grade = "A"
ElseIf score >= 80 Then
    grade = "B"
ElseIf score >= 70 Then
    grade = "C"
ElseIf score >= 60 Then
    grade = "D"
Else
    grade = "F"
End If
' Display the grade in a message box
MsgBox "The grade is: " & grade
End Sub
```

Explanation:

- **Score Evaluation**: The code evaluates the score and assigns a grade based on ranges (90 and above for A, 80–89 for B, etc.).
- **Output**: The MsgBox displays the final grade.

SUMMARY OF KEY CONCEPTS

- **If...Then**: Use for simple, single-condition checks.
- **ElseIf and Else**: Useful for evaluating multiple conditions in a single structure.
- **Select Case**: A cleaner way to handle many possible values for a single variable.
- **Logical Operators**: Combine multiple conditions using And, Or, and Not to create more complex logic.

By mastering conditional logic, you gain greater control over the flow of your VBA programs, enabling your code to respond dynamically to different inputs and conditions.

16. Cell Renaming

In MicroStation, cells are reusable groups of elements that can be named and referenced throughout your drawing. Renaming cells programmatically is useful for enforcing naming conventions or updating cells dynamically.

1. Renaming Cells Programmatically

YOU CAN RENAME CELLS in MicroStation using VBA by accessing the Name property of the CellElement. This is particularly useful when working on large projects where cells need to follow specific naming conventions.

Example: Renaming a Cell by its Element ID

VBA Code 72

```
SUB MAINPROCEDURE()
    RenameCellByID 468147, "NewName" ' give id , and new name to be given
    End Sub
    Sub RenameCellByID(elementID As Long, newCellName As String)
    Dim oCell As CellElement
    ' Retrieve the cell element by its ID
    Set oCell = ActiveModelReference.GetElementByID64(elementID)
    ' Rename the cell
    oCell.Name = newCellName
    ' Commit the changes to the model
    oCell.Rewrite
    Debug.Print "New name Given.."
    End Sub
```

Explanation:

- **GetElementByID64**: This function retrieves the cell element using its unique 64-bit elementID.
- **Renaming the Cell**: The Name property is updated with the new cell

name.
- **Saving Changes**: The Rewrite method commits the changes to the active model, ensuring that the new name is saved.

2. Best Practices for Renaming Cells

1. **Ensure Unique Names**: When renaming cells, always ensure that the new names are unique within the drawing. Duplicate names can cause confusion, especially when referencing or placing cells in different parts of the project.
2. **Consistent Naming Conventions**: Establish a standard naming convention for cells across the project to make it easier to manage and locate them. For example, use prefixes or suffixes to indicate the purpose or type of the cell.
3. **Validate Element IDs**: Always validate that the elementID exists in the active model before attempting to rename a cell. This prevents errors when working with large or complex models.

17. Working with Coordinates

In MicroStation, elements are defined in a 3D space, with each point having X, Y, and Z coordinates. You can programmatically read or modify these coordinates to control the position and shape of elements.

1. Reading Element Coordinates with VBA

WHEN WORKING WITH ELEMENTS that consist of multiple vertices (such as lines or polygons), you can extract and print the coordinates of each vertex using VBA.

Example: Reading Coordinates of a Vertex List Element

VBA Code 73

```
SUB READELEMENTCOORDINATES()
    Dim oElement As Element
    ' Retrieve the element by its ID
    Set oElement = ActiveModelReference.GetElementByID64(12345)
    ' Replace with actual ID
    ' Check if the element contains a list of vertices
    If oElement.IsVertexList Then
    Dim vertices() As Point3d
    vertices = oElement.AsVertexList.GetVertices
    ' Loop through and print each vertex's coordinates
    Dim i As Integer
    For i = LBound(vertices) To UBound(vertices)
    Debug.Print "Vertex " & i & ": (" & vertices(i).x & _
    ", " & vertices(i).y & ", " & vertices(i).Z & ")"
    Next i End If
End Sub
```

Explanation:

- **IsVertexList**: This property checks if the element contains a list of

vertices (e.g., a line or shape).
- **GetVertices**: Retrieves the coordinates of each vertex in the form of an array of Point3d.
- **Iterating Through Vertices**: A loop prints the X, Y, and Z coordinates of each vertex.

Example: Reading Cell Coordinates and Name

ANOTHER EXAMPLE IS to read cell coordinates. To read the coordinates of a cell and its name, you can use the following VBA code:

VBA Code 74

```
SUB READCELLCOORDINATESANDNAME()
    Dim oElement As element
    Dim elementID As Long
    ' Specify the element ID of the cell you want to retrieve
    elementID = 468452 ' Replace with the actual cell ID
    ' Retrieve the element by its ID
    Set oElement = ActiveModelReference.GetElementByID64(elementID)
    ' Check if the element is valid
    If oElement Is Nothing Then
    Debug.Print "Element with ID " & elementID & " not found."
    Exit Sub
    End If
    ' Check if the element is a cell element
    If oElement.IsCellElement Then
    Dim cellElement As cellElement
    Set cellElement = oElement.AsCellElement
    ' Get the cell name
    Debug.Print "Cell Name: " & cellElement.Name
    ' Get the cell's transformation (origin point)
    Dim origin As Point3d
    origin = cellElement.origin
    ' Print the origin coordinates
    Debug.Print "Cell Origin: (" & origin.x & _
    ", " & origin.y & ", " & origin.z & ")"
    Else
    Debug.Print "Element is not a cell."
```

End If
End Sub

This code reads the coordinates of a cell element and its name, helping you understand how to work with coordinates in MicroStation effectively.

2. Best Practices for Working with Coordinates

1. **Understand 3D Coordinates**: MicroStation operates in a 3D environment, so always be mindful of the Z coordinate even if you are working with 2D elements. A flat 2D element may still have a Z value that you need to account for.
2. **Coordinate Conversion**: When manipulating coordinates, ensure that the transformations (like moving or rotating elements) are handled correctly, especially when working across different views or coordinate systems.
3. **Optimized Loops**: When reading or manipulating multiple vertices, be cautious of performance in large models. Optimize loops by avoiding unnecessary operations inside them.

18.
Enumeration and Scanning Elements

Overview of Scanning Elements in MicroStation

Scanning elements in MicroStation is a crucial task that allows you to collect elements from a model based on specific criteria. This process is especially helpful when dealing with large or complex designs where you need to isolate certain elements, such as lines, circles, or custom-defined components. Scanning is typically the first step before modifying or analyzing a group of elements.

How to Scan Elements in a Drawing

TO SCAN ELEMENTS IN MicroStation, you utilize the ElementScanCriteria object to specify what types of elements you're interested in. You can filter by element type, level, color, or any other property, depending on what you need to target. Once the scan is complete, you get an ElementEnumerator object that allows you to move through the collection of scanned elements and perform actions on each one.

Example: Scanning for Line Elements

THE FOLLOWING VBA CODE demonstrates how to scan a model for line elements and print each element's ID to the immediate window:

VBA Code 75

```
SUB SCANFORLINEELEMENTS()
    ' Declare the scan criteria and set it to scan for line elements only
    Dim oCriteria As ElementScanCriteria
    Set oCriteria = New ElementScanCriteria
    oCriteria.ExcludeAllTypes
    oCriteria.IncludeOnlyVisible
    oCriteria.IncludeType msdElementTypeLine
    ' Perform the scan using the specified criteria
    Dim oEnumerator As ElementEnumerator
```

INTRODUCTION TO MICROSTATION VBA

```
Set oEnumerator = ActiveModelReference.Scan(oCriteria)
' Iterate through each element found by the scan
Do While oEnumerator.MoveNext
Dim oElement As Element
Set oElement = oEnumerator.Current
' Process each line element (e.g., print its ID)
Debug.Print "Line Element ID: " & oElement.ID64
'it fails here
Loop
End Sub
```

In this example:

- The ElementScanCriteria object is set to include only line elements (msdElementTypeLine).
- ActiveModelReference.Scan executes the scan based on the criteria.
- The ElementEnumerator is used to step through each found element. Here, it simply prints the element's ID, but in a real scenario, you might modify or further process the element.

Enumerating Through Scanned Elements

AFTER PERFORMING A scan, you'll often want to manipulate the scanned elements. The ElementEnumerator object lets you loop through each element and perform operations, such as modifying properties, moving them, or even deleting them.

Example: Modifying Scanned Elements

IN THE FOLLOWING EXAMPLE, every element in the scanned collection will have its color changed to red:

VBA Code 76

```
SUB MODIFYSCANNEDELEMENTS()
    ' Perform a scan without any specific criteria (scans all elements)
    Dim oEnumerator As ElementEnumerator
    Set oEnumerator = ActiveModelReference.Scan
```

```
' Loop through each element found in the scan
Do While oEnumerator.MoveNext
Dim oElement As Element
Set oElement = oEnumerator.Current
' this part catches all the errors and
'allow subrouting to continue
On Error Resume Next
' Change the color of the element to red (color index 1)
oElement.Color = 1
oElement.Rewrite ' Apply the changes to the element
Loop
End Sub
```

In this example:

- ActiveModelReference.Scan is used without specific criteria, meaning it will scan all elements.
- Each element's color is set to red by assigning the color index 1 (where 1 represents red in MicroStation).
- The Rewrite method is called to save the modification back into the model. Without Rewrite, the changes won't be applied.

Best Practices for Scanning and Enumerating

- **Specificity:** Always aim to be as specific as possible when defining your scan criteria. The more specific your criteria, the fewer elements will need to be processed, which can improve performance, especially in large models.

FOR EXAMPLE, IF YOU'RE only interested in line elements on a particular level or color, include these restrictions in the scan criteria:

VBA Code 77

```
OCRITERIA.INCLUDELEVEL "LevelName"
    oCriteria.IncludeColor 3
```

- **Performance Considerations:** Scanning large models can be resource-intensive. If you don't need all the elements in the model,

narrow your scan to reduce memory and processing overhead.
- **Error Handling:** Always be prepared for scenarios where no elements match your criteria. Check if the ElementEnumerator contains elements before performing actions, to avoid unnecessary errors.

By following these practices, you can ensure efficient and effective handling of elements within MicroStation models. Scanning and enumerating are core tasks in automating workflows and are key to working with large datasets in MicroStation VBA.

19.
Working with Excel

Exporting Data from MicroStation to Excel

You can export data from MicroStation elements into an Excel spreadsheet using VBA. This is useful for creating reports or further analyzing data related to the elements, such as their properties or coordinates. By utilizing the Excel Object Model in VBA, you can automate the process of writing element data to Excel, making it a smooth workflow between MicroStation and Excel.

To work with Excel from MicroStation VBA (or any VBA environment), you need to add a reference to the Microsoft Excel Object Library. This will allow you to use Excel-specific objects and methods in your code, such as Workbooks, Worksheets, and Cells. Here's a step-by-step guide on how to load the Excel library and set up the necessary references:

Steps to Load Microsoft Excel Library in VBA

1. OPEN THE VBA EDITOR

In MicroStation (or any other VBA-supported environment):

- Press Alt + F11 to open the **VBA Editor**.

2. Access References Dialog

Once in the VBA Editor:

- Go to the **Tools** menu.
- Select **References** from the dropdown menu.

3. Add Microsoft Excel Object Library

In the **References** dialog:

- Scroll through the list of available references until you find **Microsoft Excel Object Library**. The version number will depend on the version of Microsoft Office installed on your computer (e.g.,

INTRODUCTION TO MICROSTATION VBA 113

"Microsoft Excel 16.0 Object Library" for Office 2016).
- Check the box next to **Microsoft Excel Object Library**.
- Click **OK** to confirm.

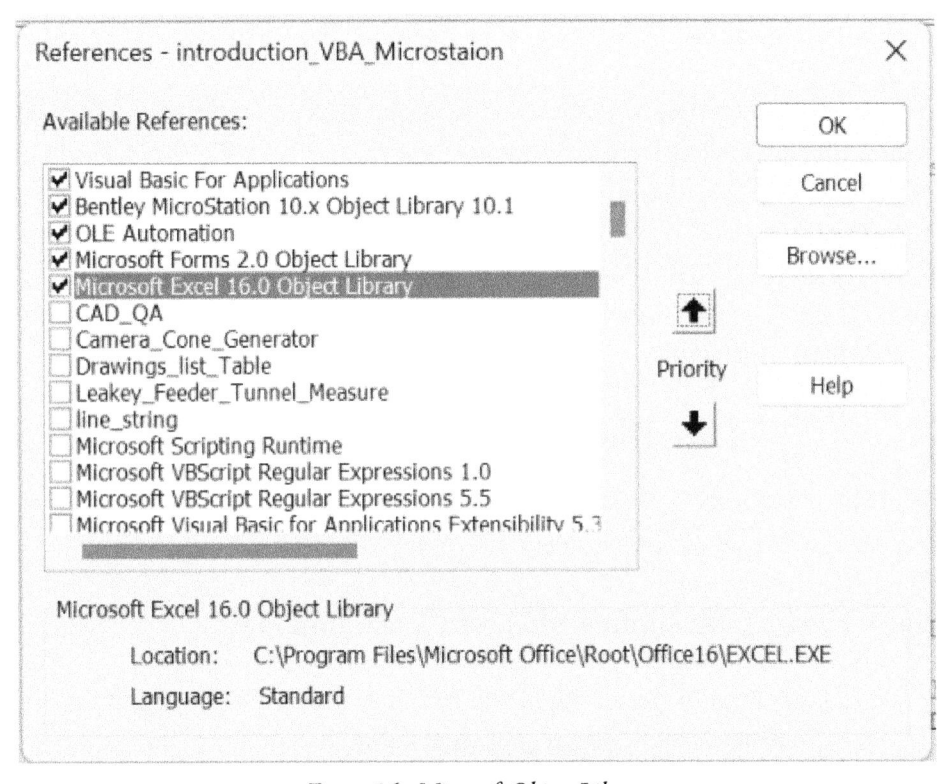

Figure 16 - Microsoft Object Library

4. VERIFY THE REFERENCE is Loaded

After adding the reference, you can verify that it's loaded by using Excel objects in your VBA code. For example, you can create an Excel application object like this:

VBA Code 78

```
SUB TESTEXCELREFERENCE()
    Dim oExcel As Excel.Application
    Set oExcel = New Excel.Application
    oExcel.Visible = True
    ' Make Excel visible to verify the library is working
```

```
' Example of adding a new workbook
Dim oWorkbook As Excel.Workbook
Set oWorkbook = oExcel.Workbooks.Add
' Clean up
oExcel.Quit
Set oExcel = Nothing
End Sub
```

This code should open xl file and close it quicky. If you want to keep it open remove the lines below from the code.

VBA Code 79

OEXCEL.QUIT
    Set oExcel = Nothing

Late Binding vs. Early Binding

WHEN WORKING WITH EXCEL or other external libraries in VBA, there are two types of binding methods:

Early Binding (Recommended for Development):

- Early binding occurs when you reference the Excel Object Library via the **References** dialog as explained above.
    - This allows you to access Excel's IntelliSense (auto-completion) while writing VBA code, making development easier.
    - You must declare your objects with their specific types, like Excel.Application, Excel.Workbook, etc.

VBA Code 80

DIM OEXCEL AS EXCEL.Application

Late Binding (No Need for Reference):

- Late binding does not require adding a reference to the Excel library.

Instead, Excel objects are created dynamically at runtime using the CreateObject function.
- This method is less strict, but it disables IntelliSense and error-checking for Excel objects during development.

VBA Code 81

```
DIM OEXCEL AS OBJECT
    Set oExcel = CreateObject("Excel.Application")
```

This approach is useful when you need to distribute your VBA project to other users who may not have the same version of Excel installed.

Key Differences Between Early and Late Binding:

- **Early Binding:**

Requires setting a reference to the Excel Object Library.
Provides IntelliSense and better performance.
More specific about the version of Excel used.

- **Late Binding:**

No need to add a reference.
Code is more flexible and works with different Excel versions.
No IntelliSense, so you'll need to manually type object members.

Summary

- Open **Tools > References** in the VBA Editor.
- Check **Microsoft Excel Object Library** to add the reference for early binding.
- Use early binding for ease of development with IntelliSense, or late binding for more flexibility when distributing your project.

ADDING THE REFERENCE to the Excel library makes it easier to automate tasks between MicroStation and Excel, providing a powerful way to handle data across the two platforms.

Example of Exporting Element Data to Excel:

NOW WE HAVE EXCEL WORKING let's see the application of Excell in Microstation. This example demonstrates how to export element data from MicroStation to an Excel sheet, such as the element's ID and its coordinates. Add a cells in your dgn file , get the ID of the cell and add this code replace the Cell then run it.

VBA Code 82

```
SUB EXPORTELEMENTDATATOEXCEL()
    ' Create an Excel application object
    Dim Element_id As Long
    Dim oExcel As Object
    Set oExcel = CreateObject("Excel.Application")
    oExcel.Visible = True ' Make Excel visible to the user
    ' Create a new workbook and worksheet
    Dim oWorkbook As Object
    Set oWorkbook = oExcel.Workbooks.Add
    Dim oWorksheet As Object
    Set oWorksheet = oWorkbook.Sheets(1)
    ' Write the headers to the Excel sheet
    oWorksheet.Cells(1, 1).value = "Element ID"
    oWorksheet.Cells(1, 2).value = "Name"
    oWorksheet.Cells(1, 3).value = "X Coordinate"
    oWorksheet.Cells(1, 4).value = "Y Coordinate"
    oWorksheet.Cells(1, 5).value = "Z Coordinate"
    ' Retrieve an element from MicroStation
    'using its ID and replace the id
    Element_id = 468452
    Dim oElement As element
    On Error Resume Next
    ' Handle potential errors in element retrieval
    Set oElement = _
    ActiveModelReference.GetElementByID(DLongFromLong(Element_id))
    On Error GoTo 0 ' Resume normal error handling
```

```
' Ensure the element was retrieved successfully
If Not oElement Is Nothing Then
' Export the element's ID and coordinates to Excel
oWorksheet.Cells(2, 1).value = oElement.ID64
oWorksheet.Cells(2, 2).value = oElement.AsCellElement.Name
oWorksheet.Cells(2, 3).value = _
oElement.AsCellElement.origin.x
oWorksheet.Cells(2, 4).value = _
oElement.AsCellElement.origin.y
oWorksheet.Cells(2, 5).value = _
oElement.AsCellElement.origin.z
Else
MsgBox "Element not found."
End If
End Sub
```

Explanation:

- **Creating an Excel Object:** The CreateObject("Excel.Application") method is used to create an instance of Excel. The workbook and worksheet objects are created, allowing you to manipulate the Excel document.
- **Exporting Element Data:** The code retrieves an element based on its ID (DLongFromLong(Element_id)) and writes its properties, such as ID and coordinates, into the appropriate cells of the Excel worksheet.
- **Error Handling:** If the element cannot be found, the program safely exits the procedure without crashing by using On Error Resume Next. A message box informs the user if the element was not found.

Importing Data from Excel into MicroStation

JUST AS YOU CAN EXPORT data from MicroStation to Excel, you can also import data from an Excel sheet to create or modify elements in MicroStation. This is especially useful when working with predefined lists of coordinates, dimensions, or other element properties that can be directly imported into the model.

Before import data from Excel use the code below to create the Data file required for the next example. VBA code below will create a new Excel file, save it to your desktop, and insert the necessary data in the format you need for the Sub ImportCoordinatesFromExcel process. This script will create a simple Excel sheet with X and Y coordinates, which can then be read and used in MicroStation to create points.

VBA Code: Create Excel File and Save It on Desktop

VBA Code 83

```
SUB CREATEEXCELFILEWITHCOORDINATES()
    ' Declare variables
    Dim oExcel As Object
    Dim oWorkbook As Object
    Dim oWorksheet As Object
    Dim desktopPath As String
    Dim savePath As String
    ' Get the path to the user's desktop
    desktopPath = _
    CreateObject("WScript.Shell").SpecialFolders("Desktop")
    savePath = desktopPath & "\CoordinatesData.xlsx"
    ' Define the full path for saving the file
    ' Create an Excel application object
    Set oExcel = CreateObject("Excel.Application")
    oExcel.Visible = True ' Optional:
    'set to False if you don't want to show Excel during the process
    ' Create a new workbook and worksheet
    Set oWorkbook = oExcel.Workbooks.Add
    Set oWorksheet = oWorkbook.Sheets(1)
    ' Add headers for the data
    oWorksheet.Cells(1, 1).Value = "X Coordinate"
    oWorksheet.Cells(1, 2).Value = "Y Coordinate"
    ' Add example data (replace this with real data as needed)
    oWorksheet.Cells(2, 1).Value = 100.5
    oWorksheet.Cells(2, 2).Value = 200.75
    oWorksheet.Cells(3, 1).Value = 150.25
    oWorksheet.Cells(3, 2).Value = 250.60
    oWorksheet.Cells(4, 1).Value = 200.1
```

```
oWorksheet.Cells(4, 2).Value = 300.9
' Save the workbook to the desktop
oWorkbook.SaveAs savePath
' Notify the user that the file has been created and saved
MsgBox "Excel file created and saved to: " & savePath
' Clean up
oWorkbook.Close
oExcel.Quit
Set oWorkbook = Nothing
Set oWorksheet = Nothing
Set oExcel = Nothing
End Sub
```

Code Explanation:

1. **Desktop Path:** The desktopPath variable stores the path to the user's desktop, and the file will be saved as CoordinatesData.xlsx.
2. **Creating Excel File:** The code creates a new Excel application, adds a workbook, and writes sample coordinates to the first few rows (you can replace this data with actual coordinates).
3. **Saving the File:** The SaveAs method is used to save the newly created Excel file to the desktop.
4. **Cleanup:** After saving the file, the workbook is closed, and the Excel application is quit to release resources.

Example of Importing Coordinates from Excel to Create Circles:

NOW WE ARE READY TO import data from Excel This code demonstrates how to read coordinates from an Excel sheet and use them to create circles in MicroStation.

VBA Code 84

```
SUB IMPORTCOORDINATESANDCREATECIRCLES()
    ' Create an Excel application object
    Dim oExcel As Object
    Set oExcel = CreateObject("Excel.Application")
```

```
desktopPath = _
CreateObject("WScript.Shell").SpecialFolders("Desktop")
filePath = desktopPath & "\CoordinatesData.xlsx"
' Define the full path for saving the file
' Open the Excel workbook containing the coordinate data
Dim oWorkbook As Object
' add path to your xl file below
Set oWorkbook = oExcel.Workbooks.Open(filePath)
Dim oWorksheet As Object
Set oWorksheet = oWorkbook.Sheets(1)
Dim x As Double, y As Double
Dim row As Integer: row = 2
' Start reading from row 2 (assuming row 1 contains headers)
' Loop through each row of the Excel sheet
'until a blank cell is encountered
Do While oWorksheet.Cells(row, 1).value <> ""
' Read the X and Y coordinates from the Excel sheet
x = oWorksheet.Cells(row, 1).value ' X coordinate
y = oWorksheet.Cells(row, 2).value ' Y coordinate
' Create a circle element in MicroStation
'at the imported coordinates
Dim oCircle As EllipseElement
Dim centerPoint As Point3d
Dim radius As Double: radius = 1#
' Set a fixed radius for the circle (adjust as needed)
' Define the center point for the circle using
'the imported X and Y coordinates
centerPoint = Point3dFromXYZ(x, y, 0)
' Create the circle element with
'the given center point and radius
Set oCircle = _
CreateEllipseElement2 _
(Nothing, centerPoint, radius, radius, Matrix3dIdentity)
' Add the circle element to the active model
ActiveModelReference.AddElement oCircle
' Move to the next row in the Excel sheet
row = row + 1
Loop
' Clean up: Close the workbook and quit Excel
oWorkbook.Close False
' False means do not save changes to the workbook
oExcel.Quit
```

```
Set oWorkbook = Nothing
Set oWorksheet = Nothing
Set oExcel = Nothing
End Sub
```

Explanation:

- **Opening an Excel Workbook:** The code opens an existing Excel file that contains the coordinate data you want to import. Each row in the sheet is expected to have X and Y coordinates.
- **Reading Data and Creating Circles:** The loop reads the X and Y values from each row of the sheet, then creates a cirlce element in MicroStation at those coordinates using the CreateEllipseElement2 method.

Best Practices for Working with Excel:

1. **Consistency in Data:** Ensure that the data in the Excel sheet is formatted consistently. For example, when importing coordinates, the cells should contain valid numbers and be free from text or other characters that could cause errors.
2. **Error Handling:** Always include error handling when dealing with Excel files. Issues such as missing files, incorrect paths, or corrupt data can crash your program. You can use On Error statements in VBA to catch and handle such errors gracefully.

Exporting Multiple Element Coordinates to Excel

WHEN WORKING WITH MULTIPLE elements, you can export their data to Excel for easier analysis. This section covers exporting the coordinates of all scanned elements from MicroStation to Excel. For this example, place few circles on the model first before using the code

Example of Exporting Coordinates of Multiple Elements to Excel:

VBA Code 85

```
SUB EXPORTCIRCLECOORDINATESTOEXCEL()
    ' Create an Excel application object
    Dim oExcel As Object
    Set oExcel = CreateObject("Excel.Application")
    oExcel.Visible = True ' Make Excel visible to the user
    ' Create a new workbook and worksheet
    Dim oWorkbook As Object
    Set oWorkbook = oExcel.Workbooks.Add
    Dim oWorksheet As Object
    Set oWorksheet = oWorkbook.Sheets(1)
    ' Write headers to the worksheet
    oWorksheet.Cells(1, 1).value = "Element ID"
    oWorksheet.Cells(1, 2).value = "X Coordinate"
    oWorksheet.Cells(1, 3).value = "Y Coordinate"
    ' Scan and enumerate through all ellipse
    'elements (circles) in the active model
    Dim oEnumerator As ElementEnumerator
    Dim oElement As Element
    Set oEnumerator = ActiveModelReference.Scan
    Dim i As Integer: i = 2 ' Row counter for Excel output
    ' Loop through elements in the model
    Do While oEnumerator.MoveNext
    Set oElement = oEnumerator.Current
    ' Check if the element is an ellipse element (circle)
    If oElement.Type = msdElementTypeEllipse Then
    Dim oEllipse As EllipseElement
    Set oEllipse = oElement.AsEllipseElement
    ' Export the element's ID and centroid coordinates
    oWorksheet.Cells(i, 1).value = oEllipse.ID64
    oWorksheet.Cells(i, 2).value = oEllipse.Centroid.x
    oWorksheet.Cells(i, 3).value = oEllipse.Centroid.y
    ' Move to the next row in Excel
    i = i + 1
    End If
    Loop
    ' Notify user that the process is complete
```

```
MsgBox "Circle coordinates have been exported to Excel."
' Clean up: Close the workbook and quit Excel
oWorkbook.SaveAs CreateObject _
("WScript.Shell").SpecialFolders("Desktop") _
& "\CircleCoordinates.xlsx"
oWorkbook.Close
oExcel.Quit
' Release objects
Set oWorksheet = Nothing
Set oWorkbook = Nothing
Set oExcel = Nothing
End Sub
```

Best Practices:

- **Organize Data in Excel:** When exporting data, organize it into well-structured tables with headers. This makes it easier to read and interpret in Excel.
- **Error Handling:** If any element does not have valid data (such as missing coordinates or ID), the error handling ensures that the process continues without crashing.

20.
Working From Excel to Microstation

To work with both Excel and MicroStation from Excel's VBA environment, you can automate MicroStation directly from Excel. This is useful if you want to run tasks in MicroStation based on data or conditions managed in Excel. Here's a step-by-step guide and code to help you achieve that.

Steps to Open Excel, Visual Basic, and Connect to MicroStation

1. Open Excel and Access the VBA Editor

STEP-BY-STEP INSTRUCTIONS:

1. Open Excel:
Start by opening Excel on your computer. You can open an existing workbook or a new blank workbook.

2. Enable the Developer Tab (if not already visible):
The Developer tab gives you access to the VBA Editor and other advanced Excel features. If it's not visible, follow these steps:

1. Click on the **File** menu in Excel.
2. Select **Options** at the bottom of the menu.
3. In the **Excel Options** dialog box, select **Customize Ribbon** from the left-hand side.
4. On the right-hand side, check the box for **Developer** under the "Main Tabs" section.
5. Click **OK** to close the dialog box. You should now see the **Developer** tab on the Excel ribbon.

3. Open the VBA Editor:
Once the Developer tab is visible, follow these steps:

1. Click on the **Developer** tab in the Excel ribbon.

INTRODUCTION TO MICROSTATION VBA 125

2. In the **Code** group, click on the **Visual Basic** button. Alternatively, you can press the keyboard shortcut **Alt + F11** to open the VBA Editor directly.

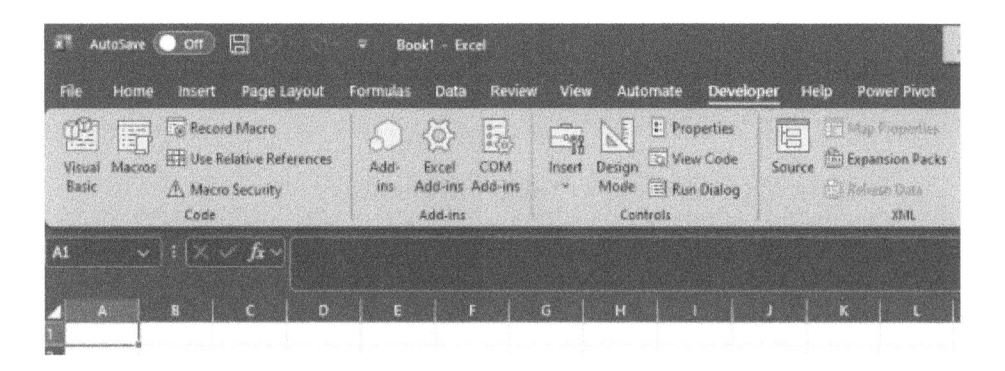

Figure 17 - Excel VBA Editor

4. EXPLORE THE VBA Editor:

The VBA Editor will open in a new window. You'll see various components, including:

1. **Project Explorer**: Displays all the workbooks and worksheets that are currently open.
2. **Code Window**: This is where you'll write and edit your VBA code.
3. **Properties Window**: Allows you to view and modify properties of selected objects.

5. Insert a New Module (optional):

To start writing VBA code:

1. In the VBA Editor, go to the **Insert** menu.
2. Select **Module** from the dropdown.
3. A new blank code window will appear, ready for you to start writing your VBA code.

6. Return to Excel:

To go back to Excel from the VBA Editor, simply click the **X** to close the VBA Editor window or press **Alt + Q**.

7. Keyboard Shortcut for Quick Access:

Alt + F11: This is the fastest way to toggle the VBA Editor on and off while working in Excel.

2. Add a Reference to MicroStation

TO INTERACT WITH MICROSTATION from Excel, you need to add a reference to the MicroStation library:

- In the VBA editor, go to **Tools > References**.
- Scroll down to find **Bentley MicroStation xx object Library xx.x** (where xx represents the version of MicroStation you are using).
- Check the box and click **OK**.

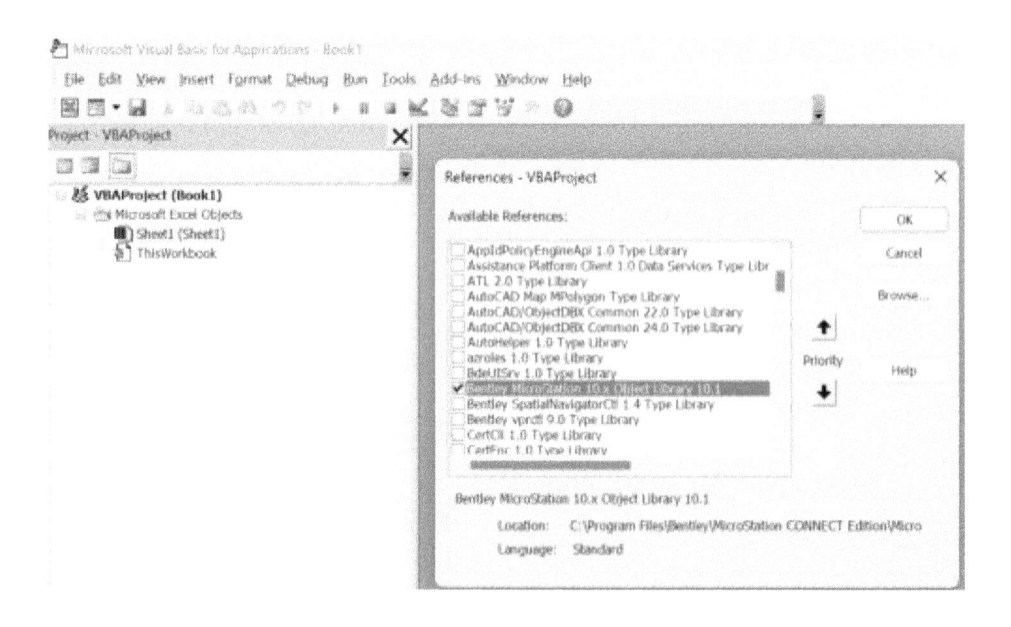

Figure 18 - Microstation Object Library

3. Example Code to Connect Excel to MicroStation

BELOW IS A VBA EXAMPLE in Excel that connects to MicroStation, retrieves data from Excel (like coordinates), and places elements in MicroStation based on the Excel data.

Example 1: Connect to MicroStation and Place a Line

Enter the Coordinates for the Start of the Line:

- The start coordinates of the line will go in **row 1** (below the labels):
    - In **cell A1**, enter the **X** coordinate of the start point (e.g., 0).
    - In **cell B1**, enter the **Y** coordinate of the start point (e.g., 0).
    - In **cell C1**, enter the **Z** coordinate of the start point (e.g., 0).

Enter the Coordinates for the End of the Line:

- The end coordinates of the line will go in **row 2**:
    - In **cell A2**, enter the **X** coordinate of the end point (e.g., 5).
    - In **cell B2**, enter the **Y** coordinate of the end point (e.g., 5).
    - In **cell C2**, enter the **Z** coordinate of the end point (e.g., 0).

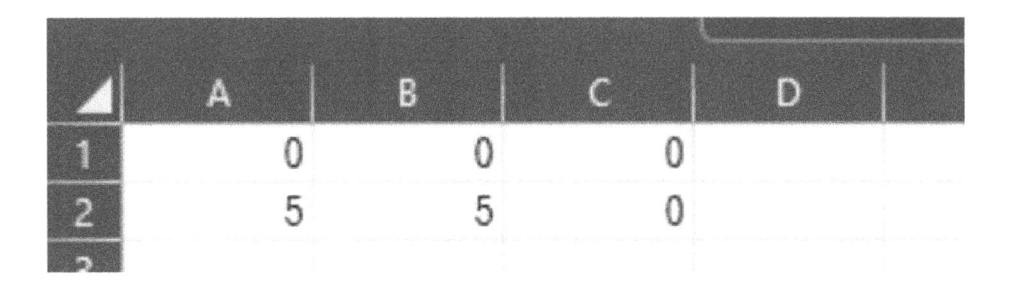

	A	B	C	D
1	0	0	0	
2	5	5	0	

Figure 19 - Excel Data

Explanation of the Data:

- **Row 2**: Contains the coordinates for the **start point** of the line.
- **Row 3**: Contains the coordinates for the **end point** of the line.

IN THIS SETUP:

- **X** represents the position along the horizontal axis.
- **Y** represents the position along the vertical axis.

- **Z** represents the depth or height in 3D space.

Once coordinates have been entered on Excel, Open VBA Editor in Excel and enter the code below. This code connects Excel to MicroStation and creates a line based on coordinates stored in Excel.

VBA Code 86

```
SUB CONNECTTOMICROSTATIONANDPLACELINE()
    ' Declare MicroStation application and objects
    Dim oMicroStation As Object
    Dim oLine As LineElement
    Dim startPoint As Point3d, endPoint As Point3d
    ' Create a connection to MicroStation
    On Error Resume Next
    Set oMicroStation = GetObject(, "MicroStationDGN.Application") ' Connect to an open instance of MicroStation
    If oMicroStation Is Nothing Then
    MsgBox "MicroStation is not running."
    Exit Sub
    End If
    On Error GoTo 0
    ' Make MicroStation visible (optional)
    oMicroStation.Visible = True
    ' Define the coordinates (you can get these from Excel cells)
    ' Let's assume the coordinates are in cells
    'A1 (start X), B1 (start Y), A2 (end X), B2 (end Y)
    startPoint.X = Range("A1").Value
    startPoint.Y = Range("B1").Value
    startPoint.Z = 0 ' Set Z to 0 for a 2D line
    endPoint.X = Range("A2").Value
    endPoint.Y = Range("B2").Value
    endPoint.Z = 0 ' Set Z to 0 for a 2D line
    ' Create a line element in MicroStation
    Set oLine = oMicroStation.CreateLineElement2 _
    (Nothing, startPoint, endPoint)
    ' Add the line element to the active MicroStation model
    oMicroStation.ActiveModelReference.AddElement oLine
    MsgBox "Line has been placed in MicroStation."
    ' Clean up
```

```
Set oLine = Nothing
Set oMicroStation = Nothing
End Sub
```

4. Breakdown of the Code:

1. CONNECTING TO MICROSTATION:

- The GetObject(, "MicroStationDGN.Application") function is used to connect to an active instance of MicroStation. If MicroStation is not open, it displays a message.

2. Accessing Excel Data:

- The code reads values from specific Excel cells (e.g., A1, B1, A2, and B2) as the start and end coordinates for the line.

3. Creating Elements in MicroStation:

- The CreateLineElement2 function is used to create a line in MicroStation between two points. This line is added to the active model using ActiveModelReference.AddElement.

4. Error Handling:

- Basic error handling is included to manage cases where MicroStation is not running.

5. Example 2: Reading Multiple Coordinates from Excel and Placing Circles in MicroStation

THIS CODE READS MULTIPLE rows of X and Y coordinates from Excel and places circles in MicroStation at those locations.

VBA Code 87

```
SUB PLACECIRCLESFROMEXCEL()
    ' Declare MicroStation application and objects
    Dim oMicroStation As Object
    Dim oCircle As EllipseElement
    Dim centerPoint As Point3d
    Dim radius As Double: radius = 1 ' Fixed radius for all circles
    ' Create a connection to MicroStation
    On Error Resume Next
    Set oMicroStation = GetObject(, "MicroStationDGN.Application")
    If oMicroStation Is Nothing Then
    MsgBox "MicroStation is not running."
    Exit Sub
    End If
    On Error GoTo 0
    ' Make MicroStation visible (optional)
    oMicroStation.Visible = True
    ' Set the starting row
    ' (assumes coordinates start from row 2, with headers in row 1)
    Dim row As Integer: row = 2
    ' Loop through rows in Excel until an
    'empty cell in column A is encountered
    Do While Range("A" & row).Value <> ""
    ' Read the X and Y coordinates from the Excel sheet
    centerPoint.X = Range("A" & row).Value
    centerPoint.Y = Range("B" & row).Value
    centerPoint.Z = 0 ' Set Z to 0 for 2D circles
    ' Create a circle element in MicroStation
    Set oCircle = _
    oMicroStation.CreateEllipseElement2 _
    (Nothing, centerPoint, radius, radius, Matrix3dIdentity)
    ' Add the circle element to the active MicroStation model
    oMicroStation.ActiveModelReference.AddElement oCircle
    ' Move to the next row
    row = row + 1
    Loop
    MsgBox "Circles have been placed in MicroStation."
    ' Clean up
    Set oCircle = Nothing
```

```
Set oMicroStation = Nothing
End Sub
```

6. Best Practices:

- **Error Handling:** Always include error-handling mechanisms to ensure that the connection to MicroStation is stable and that no runtime errors cause the program to fail.
- **Data Validation:** Validate the Excel data to ensure that the coordinates are numeric and properly formatted.
- **Closing MicroStation Objects:** Always set objects to Nothing after use to prevent memory leaks.

Exercise: Working with Text IDs and Notes in Excel Using VBA

Problem Statement:

CREATE A VBA SOLUTION in Excel that reads **Text IDs** and **Text Notes** from a worksheet and allows users to replace notes through an interface. The worksheet will have the following structure:

- **Column 1**: Text IDs
- **Column 2**: Text Level
- **Column 3**: Text Notes
- **Column 4**: Replaced Notes (if any)

You will implement two buttons:

- **Button 1 (Read):** Reads the Text IDs and Text Notes and populates them in the worksheet.
- **Button 2 (Replace):** Replaces the original text notes with new ones entered by the user in column 3.

Step-by-Step Instructions:

1. SET UP YOUR EXCEL Sheet:

- In **Sheet1**, label **Column A** as "Text ID", **Column B** as "Levels", **Column C** as "Text Notes", and **Column D** as "New Notes".
- Example layout:

Text ID	Level	Text Notes	New Notes
101	text	Initial Note 1	(user input)
102	text	Initial Note 2	(user input)

2. Add Buttons:
Insert two buttons:

- **Button 1**: Label it "Read ".
- **Button 2**: Label it "Replace ".

To add buttons, go to **Developer Tab > Insert > Form Controls > Button**.

VBA Code Solution:

THIS VBA SOLUTION CONSISTS of four subroutines that work together to read and replace text elements in MicroStation, using data entered in **Sheet1** of Excel. Here's a breakdown of each subroutine:

1. MainProcedure

THE MAINPROCEDURE ACTS as the main function, calling other subroutines in sequence:

- It first calls Add_titles to set up the Excel sheet with titles and formatting.
- Then it calls GettextData to scan and import text data from MicroStation into the Excel sheet.

VBA Code 88

```
SUB MAINPROCEDURE()
    ' Calling other subroutines within MainProcedure
    Add_titles
    GettextData
    End Sub
```

2. Add_titles

THIS SUBROUTINE SETS up the formatting and titles for **Sheet1**:

- **Row 1** is colored yellow for the buttons to be placed later.
- **Row 2** is colored green and labeled with titles for the data:
    - **Column A**: ID
    - **Column B**: Level
    - **Column C**: Current Text
    - **Column D**: New Text (to be replaced if needed).

VBA Code 89

```
SUB ADD_TITLES()
    ' Move titles to Row 2
    Sheets("Sheet1").Cells(2, 1).Value = "ID"
    Sheets("Sheet1").Cells(2, 2).Value = "Level"
    Sheets("Sheet1").Cells(2, 3).Value = "Current Text"
    Sheets("Sheet1").Cells(2, 4).Value = "New Text"
    ' Color Row 1 Yellow
    Sheets("Sheet1").Range("A1:D1").Interior.Color = _
    RGB(255, 255, 0) ' Yellow
    ' Color Row 2 Green
    Sheets("Sheet1").Range("A2:D2").Interior.Color = _
    RGB(0, 255, 0) ' Green
    End Sub
```

3. GettextData

THIS SUBROUTINE CONNECTS to MicroStation and scans for text elements. It retrieves the text element IDs, levels, and the current text from the active model and populates the data in **Sheet1** starting from **Row 3**.

- **Column A**: ID (text element ID).
- **Column B**: Level (the level of the text element in MicroStation).
- **Column C**: Current Text (text currently present in MicroStation).
- **Column D**: New Text (pre-filled with the current text for the user to edit).

VBA Code 90

```vba
SUB GETTEXTDATA()
    ' Declare variables
    Dim mytext As TextElement
    Dim MyScanCriteria As ElementScanCriteria
    Dim myScanEnum As ElementEnumerator
    Dim oMicroStation As Object
    Dim i As Long
    ' Create a connection to MicroStation
    On Error Resume Next
    Set oMicroStation = GetObject(, "MicroStationDGN.Application") ' Connect to an open instance of MicroStation
    If oMicroStation Is Nothing Then
    MsgBox "MicroStation is not running."
    Exit Sub
    End If
    On Error GoTo 0
    ' Set up scan criteria to include only text elements
    Set MyScanCriteria = New ElementScanCriteria
    MyScanCriteria.ExcludeAllTypes
    MyScanCriteria.IncludeType msdElementTypeText
    ' Clear previous data in the worksheet range A3 to E100
    ThisWorkbook.Sheets("Sheet1").Range("A3:E100").Clear
    ' Scan for text elements in the active model
    Set myScanEnum = _
    oMicroStation.ActiveModelReference.Scan(MyScanCriteria)
    i = 3 ' Start at row 3 in the worksheet
    ' Loop through each text element found
    Do While myScanEnum.MoveNext
    Set mytext = myScanEnum.Current
    ' Populate the worksheet with text element data
    Sheets("Sheet1").Cells(i, 1).Value = mytext.ID64
    Sheets("Sheet1").Cells(i, 2).Value = mytext.Level.Name
    Sheets("Sheet1").Cells(i, 3).Value = mytext.Text
    Sheets("Sheet1").Cells(i, 4).Value = mytext.Text
    i = i + 1 ' Move to the next row
    Loop
    End Sub
```

4. Multitextreplace

THIS SUBROUTINE CHECKS **Sheet1** for new text in **Column D** and replaces the old text in MicroStation if the user has entered a replacement.

- The subroutine scans the selected text elements in MicroStation by their ID (from **Column A**).
- If a replacement is found in **Column D**, it updates the text in MicroStation.

After performing the replacements, it calls GettextData again to refresh the data in Excel.

VBA Code 91

```
SUB MULTITEXTREPLACE()
    ' Declare variables
    Dim myMSAppCon As MicroStationDGN.ApplicationObjectConnector
    Dim myMSApp As MicroStationDGN.Application
    Dim mytext As TextElement
    Dim myelem As Element
    Dim mymod As MicroStationDGN.ModelReference
    Dim myelems As ElementEnumerator
    Dim c As Range
    Dim lr As Long
    ' Find the last populated row in column A
    lr = Worksheets("Sheet1").Cells(Rows.Count, 1).End(xlUp).Row
    ' Connect to MicroStation
    Set myMSAppCon = _
    GetObject(, "MicroStationDGN.ApplicationObjectConnector")
    Set myMSApp = myMSAppCon.Application
    ' Select all elements in MicroStation
    CadInputQueue.SendCommand "CHOOSE ALL"
    ' Loop through the rows in Excel from row 3 to the last row
    For Each c In Range(Cells(3, 1), Cells(lr, 1))
        ' Select element by ID from column A
        cellid = "select byelemid " & _
        Sheets("Sheet1").Cells(c.Row, 1).Value
        myMSAppCon.Application.CadInputQueue.SendKeyin cellid
```

```
' Get selected elements
Set mymod = _
MicroStationDGN.Application.ActiveModelReference
Set myelems = mymod.GetSelectedElements
' Loop through the selected elements
While myelems.MoveNext
Set myelem = myelems.Current
Select Case myelem.Type
Case msdElementTypeText
Set mytext = myelem
' Get original and new text from columns C and D
Dim oldText As String
Dim newText As String
oldText = _
Worksheets("Sheet1").Cells(c.Row, 3).Value
newText = _
Worksheets("Sheet1").Cells(c.Row, 4).Value
' Replace the text if it has changed
If InStr(1, mytext.Text, oldText) > 0 Then
mytext.Text = _
Replace(mytext.Text, oldText, newText)
mytext.Rewrite ' Apply changes
End If
End Select
Wend
Next c
' Refresh data after replacement
GettextData
End Sub
```

How to Add Buttons in Excel to Run These Macros

To run the main procedure and the text replacement code using buttons:

Step-by-Step Instructions:

1. Insert a Button:

- Go to the **Developer** tab in Excel.
- Click **Insert > Button (Form Control)**.
- Draw a button on the sheet where you want it to appear.

2. Assign a Macro to the Button:

- After drawing the button, Excel will prompt you to assign a macro.
- For the first button, select the MainProcedure macro and click **OK**.
- Label the button as **"Read"**.

3. Add a Second Button:

- Insert another button using the same process.
- Assign the Multitextreplace macro to this button.
- Label this button as **"Replace"**.

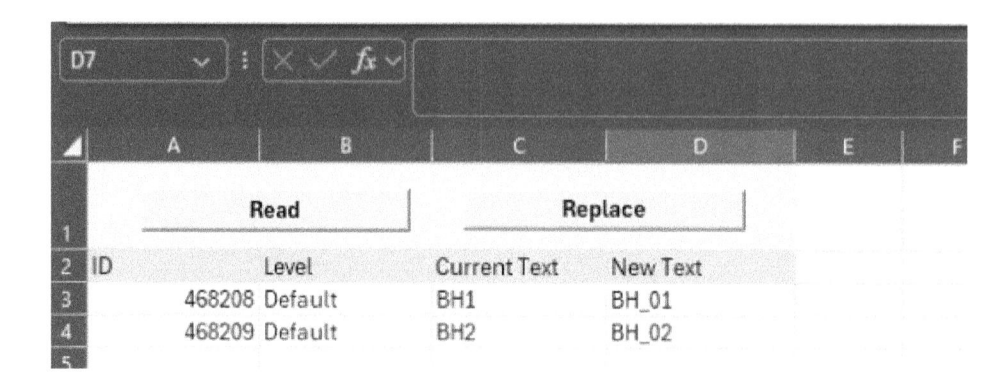

Figure 20 - Excel Text Replacement Tool

THE CODE SOLUTION I used two approaches, **late binding** and **Early Binding** methods to connect to MicroStation, which dynamically links to an existing instance of the application, this concept was explained earlier. Here's how it works:

1. Late Binding (GetObject Method)

IN THE GETTEXTDATA subroutine, the GetObject function connects to a running instance of MicroStation.

VBA Code 92

```
DIM OMICROSTATION AS Object
    Set oMicroStation = GetObject(, "MicroStationDGN.Application") ' Late binding
```

- **Late Binding**: No need for a specific library reference. MicroStation is treated as a generic Object, making it more flexible but with no IntelliSense support.

2. Early Binding (Library Reference)

WITH **early binding**, you explicitly reference the MicroStation library, allowing you to use IntelliSense and get compile-time error checking. Here's an example:

1. **Set a Reference**: In the VBA editor, go to **Tools > References** and check **"MicroStationDGN Object Library"**.
2. **Use Early Binding in Code**:

VBA Code 93

DIM OMICROSTATION AS MicroStationDGN.Application
    Set oMicroStation = New MicroStationDGN.Application ' Early binding

Benefits of Each Method

- **Late Binding** (GetObject):
    - **Flexible**: No need for library references, works across different MicroStation versions.
    - **Portable**: Easier to distribute, no version dependencies.
- **Early Binding** (Library Reference):
    - **IntelliSense**: Helps you see methods and properties while coding.
    - **Faster**: Slight performance boost and compile-time error checking.

USE **late binding** when flexibility is key and **early binding** for easier development and better performance.

summary:

YOU NOW HAVE A FULLY functional solution where one button loads the text data from MicroStation into **Sheet1**, and another button replaces text based on the user's input in **Column D** of **Sheet1**.

Conclusion

BY WRITING VBA CODE in Excel, you can automate interactions with MicroStation. This integration allows you to use Excel as a data source to drive element creation in MicroStation, such as placing lines or circles based on predefined coordinates in Excel. This method is particularly useful for automating repetitive tasks or managing large datasets.

21.
Using Excel as a Database

Leveraging Excel for Storing and Managing Data

Excel can serve as a lightweight database for storing important data, such as element attributes, coordinates, or configuration settings for your MicroStation VBA projects. This allows you to interact with the data dynamically, retrieving and updating it without the need for a more complex database system. Excel's structure of rows and columns makes it ideal for organizing and accessing data as records and fields.

Reading Data from Excel

ONE OF THE MOST COMMON operations is reading data from Excel into MicroStation. In this example, each row in Excel is treated as a record, and each column represents a field of data (e.g., coordinates, attributes).

Example of Reading Data from Excel:

VBA Code 94

```
SUB READEXCELDATAASDATABASE()
    ' Create an Excel application object
    Dim oExcel As Object
    Set oExcel = CreateObject("Excel.Application")
    ' Open an existing Excel workbook
    Dim oWorkbook As Object
    Set oWorkbook = _
    oExcel.Workbooks.Open("C:\path\to\your\file.xlsx")
    Dim oWorksheet As Object
    Set oWorksheet = oWorkbook.Sheets(1) ' Access the first sheet
    ' Initialize the row number to start
    'reading from (assumes headers in row 1)
    Dim row As Integer: row = 2
    ' Loop through each row in the Excel
    'sheet until a blank cell is encountered
```

```
Do While oWorksheet.Cells(row, 1).Value <> ""
' Output the data from each row as a "record"
On Error Resume Next ' Handle potential errors
Debug.Print "Record " & row & ": " & _
oWorksheet.Cells(row, 1).Value & ", " _
& oWorksheet.Cells(row, 2).Value
On Error GoTo 0 ' Reset normal error handling
row = row + 1 ' Move to the next row
Loop
'close the workbook
oWorkbook.Close
oExcel.Quit ' Quit Excel application
End Sub
```

Explanation:

- **Opening an Excel Workbook:** The CreateObject("Excel.Application") method opens an instance of Excel, and the workbook is opened using its file path.
- **Reading Data by Rows:** A loop iterates through each row of data, starting from row 2 (assuming row 1 contains headers). The loop continues until it encounters a blank cell in the first column, which signals the end of the data.
- **Error Handling:** The On Error Resume Next statement ensures that any errors encountered (such as reading empty cells or invalid data) are handled without crashing the program.

Updating Data in Excel

IN ADDITION TO READING data from Excel, you can also update Excel cells programmatically, treating Excel as a simple database where records can be modified. This is helpful for tasks like updating element attributes or status information from MicroStation back into an Excel sheet.

Example of Updating Data in Excel:

VBA Code 95

```
SUB UPDATEEXCELDATA()
    ' Create an Excel application object
    Dim oExcel As Object
    Set oExcel = CreateObject("Excel.Application")
    ' Open an existing Excel workbook
    Dim oWorkbook As Object
    Set oWorkbook = _
    oExcel.Workbooks.Open("C:\path\to\your\file.xlsx")
    Dim oWorksheet As Object
    Set oWorksheet = oWorkbook.Sheets(1) ' Access the first sheet
    ' Update data in specific cells
    On Error Resume Next ' Handle potential errors
    oWorksheet.Cells(2, 1).Value = "Updated Element ID"
    oWorksheet.Cells(2, 2).Value = 100 ' New X coordinate value
    oWorksheet.Cells(2, 3).Value = 200 ' New Y coordinate value
    On Error GoTo 0 ' Reset normal error handling
    ' Save and close the workbook
    oWorkbook.Save
    oWorkbook.Close
    oExcel.Quit ' Quit Excel application
End Sub
```

Explanation:

- **Updating Data:** This example demonstrates how to update the values in specific cells of an Excel worksheet. The changes are saved, and the workbook is closed after updating the data.
- **Error Handling:** Again, On Error Resume Next is used to ensure that any issues encountered while updating do not interrupt the execution of the program.

Best Practices for Using Excel as a Database

1. DATA STRUCTURE AND Organization:

- Keep your data structured with clear headers and consistent formatting. For example, when working with coordinates, ensure all relevant fields (e.g., X and Y coordinates) are always populated and in the correct columns.
- Organize your data in a table-like format for easy reading and updating. Each row should represent one complete record (e.g., an element's attributes or coordinates).

2. Data Validation in Excel:

- Use Excel's built-in data validation features to ensure the integrity of your data before importing it into MicroStation. For instance, you can restrict certain columns to accept only numbers or specific text formats, minimizing errors during import.

3. Error Handling in VBA:

- Always implement error-handling routines in your VBA code when interacting with Excel. Opening files, reading data, and writing updates can sometimes result in errors (e.g., missing files, invalid paths, or empty cells). Proper error handling will help your program run smoothly and avoid crashing unexpectedly.

4. Saving and Closing Excel Properly:

- Always remember to save your Excel workbook after making updates, and properly close both the workbook and the Excel application to prevent resource leaks.

By treating Excel as a database, you can extend its capabilities to store, retrieve, and manage data dynamically in your MicroStation VBA projects without relying on more complex database systems.

Producing a Table From Excel to MicroStation

ANOTHER GOOD CASE FOR Microstation to Excel connectivity is the Transferring data from Excel to MicroStation. This is a powerful way to automate the creation of tables in your drawings. This can significantly streamline your workflow, especially when you have tabular data that needs to be presented visually in your designs.

Purpose of the Process

THE MAIN PURPOSE OF this process is to utilize the data stored in an Excel spreadsheet to create a formatted table directly within a MicroStation drawing. By leveraging Visual Basic for Applications (VBA), you can automate the entire process, ensuring that the data is accurately represented in your design files without the need for manual entry.

Steps Involved

1. PREPARE YOUR EXCEL Spreadsheet:

- Create a worksheet in Excel where you will enter your data. For this example, ensure you prepare the cells starting from A1. Each column will represent a different piece of information, and each row will represent a different entry or record.

2. Prepare A1 Drawing Sheet:

- Before running the VBA code, make sure your MicroStation drawing sheet is set up to A1 size for this example and ensure you have a clear area where the table will be placed.

3. Write the VBA Code:
The VBA code will handle the following tasks:

- **Connect to Excel**: Establish a connection to the currently open instance of Excel to access the data.
- **Retrieve Data**: Loop through the Excel worksheet to gather the data

needed for the table.
- **Calculate Dimensions**: Determine the size and layout of the table based on the data retrieved (e.g., column widths, row heights).
- **Create Text Elements**: For each piece of data, the code will create text elements in MicroStation, placing them in the appropriate positions.
- **Draw Borders**: The code will also create line elements to visually define the table's cells.

4. Run the Macro:

- Execute the VBA macro in MicroStation. The code will read the data from the specified Excel sheet, create the table based on the data, and render it directly onto the drawing sheet.

VBA Code 96

```
SUB ADD_TABLE_MACRO()
    ' Define variables
    Dim mytext As TextElement
    Dim rotmatrix As Matrix3d
    Dim myline As LineElement
    Dim LineSt As Point3d
    Dim LineEn As Point3d
    Dim x As Integer
    Dim i As Integer
    Dim lr As Integer
    Dim maxTextLength As Integer
    Dim myExcel As Excel.Application
    Dim mysheet As Worksheet
    Dim cellText As String
    Dim colWidth() As Double
    Dim columnCount As Integer
    Dim colIndex As Integer
    Dim cumulativeX As Double
    Dim totalTableWidth As Double
    Dim totalTableHeight As Double
    Dim totalWidth As Double
```

```vba
Dim totalHeight As Double
Dim paddingLeft As Double
Dim paddingVertical As Double
' Set Excel objects
Set myExcel = GetObject(, "Excel.Application")
Set mysheet = myExcel.ActiveSheet
' Total available dimensions (height and width)
'to center the table within
totalWidth = 0.841
' Example total available width (could be user-defined)
totalHeight = 0.7088
' Example total available height (could be user-defined)
' Set cell height and text size
Dim cellHeight As Double
Dim textSize As Double
cellHeight = 0.007
' Standard cell height (could be user-defined)
textSize = 0.0025
' Fixed text size (could be user-defined)
' Padding for text inside the cell
paddingLeft = 0.001 ' Left padding
paddingVertical = -0.001 ' Vertical padding
' Find the last used row in Excel
lr = mysheet.Cells(1000, 1).End(xlUp).row + 1
' Determine the number of columns used in the first row
columnCount = mysheet.Cells(1, 1).End(xlToRight).Column
' Initialize column widths array
ReDim colWidth(1 To columnCount)
' Calculate column widths based on the
'longest text in each column
totalTableWidth = 0 ' Initialize total table width
For colIndex = 1 To columnCount
maxTextLength = 0
' Loop through the rows to find the
'longest text in the current column
For x = 1 To lr - 1
cellText = CStr(mysheet.Cells(x, colIndex).value)
If Len(cellText) > maxTextLength Then
maxTextLength = Len(cellText)
End If
Next x
' Set column width based on the longest text
```

```
colWidth(colIndex) = _
maxTextLength * textSize * 1.2
' Adjust this multiplier as needed for extra padding
' Add column width to the total table width
totalTableWidth = _
totalTableWidth + colWidth(colIndex)
Next colIndex
' Calculate the total height of the table
totalTableHeight = (lr - 1) * cellHeight
' Calculate starting X and Y positions to center the table
Dim tableStartX As Double
Dim tableStartY As Double
tableStartX = (totalWidth - totalTableWidth) / 2
' Center horizontally
tableStartY = (totalHeight + totalTableHeight) / 2
' Center vertically
' (positive because Y decreases downward in MicroStation)
' Initialize cumulativeX with tableStartX
'to begin positioning columns from there
cumulativeX = tableStartX
' Column loop for all columns in the sheet
On Error Resume Next ' Ignore any errors that may occur
For colIndex = 1 To columnCount
' Row loop
For x = 1 To lr - 1
' Get cell value as text
cellText = CStr(mysheet.Cells(x, colIndex).value)
' If it's the first row (title row),
'convert text to uppercase
If x = 1 Then
cellText = UCase(cellText) ' Convert to uppercase
End If
' Create text element
Set mytext = Application.CreateTextElement1 _
(Nothing, cellText, Point3dFromXYZ _
(cumulativeX + paddingLeft, _
tableStartY - (x - 1) * cellHeight - _
(cellHeight / 2) + paddingVertical, 0), rotmatrix)
ActiveModelReference.AddElement mytext
' Create table borders (box for each cell)
' Top line
LineSt.x = cumulativeX
```

```
LineSt.y = tableStartY - (x - 1) * cellHeight
LineSt.z = 0
LineEn.x = cumulativeX + colWidth(colIndex)
LineEn.y = tableStartY - (x - 1) * cellHeight
LineEn.z = 0
Set myline = _
Application.CreateLineElement2(Nothing, LineSt, LineEn)
ActiveModelReference.AddElement myline
' Bottom line
LineSt.x = cumulativeX
LineSt.y = tableStartY - x * cellHeight
LineSt.z = 0
LineEn.x = cumulativeX + colWidth(colIndex)
LineEn.y = tableStartY - x * cellHeight
LineEn.z = 0
Set myline = _
Application.CreateLineElement2(Nothing, LineSt, LineEn)
ActiveModelReference.AddElement myline
' Right line
LineSt.x = cumulativeX + colWidth(colIndex)
LineSt.y = tableStartY - (x - 1) * cellHeight
LineSt.z = 0
LineEn.x = cumulativeX + colWidth(colIndex)
LineEn.y = tableStartY - x * cellHeight
LineEn.z = 0
Set myline = _
Application.CreateLineElement2(Nothing, LineSt, LineEn)
ActiveModelReference.AddElement myline
' Left line
LineSt.x = cumulativeX
LineSt.y = tableStartY - (x - 1) * cellHeight
LineSt.z = 0
LineEn.x = cumulativeX
LineEn.y = tableStartY - x * cellHeight
LineEn.z = 0
Set myline = _
Application.CreateLineElement2(Nothing, LineSt, LineEn)
ActiveModelReference.AddElement myline
Next x
' After finishing with the current column,
'add its width to cumulativeX for the next column
cumulativeX = cumulativeX + colWidth(colIndex)
```

Next colIndex
On Error GoTo 0 ' Stop ignoring errors after table creation
End Sub

Code Breakdown and Explanation

VBA Code 97

SUB ADD_TABLE_MACRO()

- **Sub Declaration**: This line begins the definition of the **Add_table_Macro** subroutine, which will contain all the code to create a table in MicroStation.

Variable Declarations

VBA Code 98

```
' DEFINE VARIABLES
    Dim mytext As TextElement
    Dim rotmatrix As Matrix3d
    Dim myline As LineElement
    Dim LineSt As Point3d
    Dim LineEn As Point3d
    Dim x As Integer
    Dim i As Integer
    Dim lr As Integer
    Dim maxTextLength As Integer
    Dim myExcel As Excel.Application
    Dim mysheet As Worksheet
    Dim cellText As String
    Dim colWidth() As Double
    Dim columnCount As Integer
    Dim colIndex As Integer
    Dim cumulativeX As Double
    Dim totalTableWidth As Double
    Dim totalTableHeight As Double
    Dim totalWidth As Double
    Dim totalHeight As Double
    Dim paddingLeft As Double
    Dim paddingVertical As Double
```

- **Variable Definitions**: Here, various variables are defined to be used throughout the procedure. These include:

TextElement: Represents text elements to be created in MicroStation.

Matrix3d: Holds transformation matrices for positioning elements.

LineElement: Represents line segments for drawing table borders.

Point3d: Represents points in 3D space for line start and end positions.

Integer Variables: For counting iterations and storing row/column information.

Excel Objects: myExcel and mysheet will refer to the active Excel application and sheet.

Dynamic Arrays: colWidth() will hold the widths of each column dynamically based on content.

Cumulative Widths and Heights: Used for calculating the layout of the table.

Setting Up Excel Objects

VBA Code 99

```
' SET EXCEL OBJECTS
    Set myExcel = GetObject(, "Excel.Application")
    Set mysheet = myExcel.ActiveSheet
```

- **Excel Object Initialization**: This section connects the VBA script to the currently running instance of Excel and gets the active worksheet. This allows the macro to read data from Excel.

Table Dimensions and Padding

VBA Code 100

```
' TOTAL AVAILABLE DIMENSIONS (height and width)
    'to center the table within
    totalWidth = 0.841
    ' Example total available width (could be user-defined)
    totalHeight = 0.7088
    ' Example total available height (could be user-defined)
    ' Set cell height and text size
    Dim cellHeight As Double
    Dim textSize As Double
    cellHeight = 0.007
    ' Standard cell height (could be user-defined)
    textSize = 0.0025
    ' Fixed text size (could be user-defined)
    ' Padding for text inside the cell
    paddingLeft = 0.001 ' Left padding
    paddingVertical = -0.001 ' Vertical padding
```

- **Table Dimensions**: The total dimensions for the table are defined. These can be customized according to your layout needs.
- **Cell Height and Text Size**: These values define how tall each cell will be and the size of the text that will be placed inside each cell.
- **Padding**: This section sets how much space will be around the text within each cell to ensure it doesn't touch the borders.

Finding Last Used Row

VBA Code 101

```
' FIND THE LAST USED row in Excel
    lr = mysheet.Cells(1000, 1).End(xlUp).Row + 1
```

- **Finding Last Row**: This line determines the last row that contains data in the first column (A). It uses a common Excel trick to find the last used row by starting from a cell far down in the column and moving up.

Determine Column Count

VBA Code 102

```
' DETERMINE THE NUMBER of columns used in the first row
    columnCount = mysheet.Cells(1, 1).End(xlToRight).Column
```

- **Counting Columns**: This line counts the total number of columns that have data in the first row, which will be used to determine how many columns to include in the table.

Initialize Column Widths Array

VBA Code 103

```
' INITIALIZE COLUMN widths array
    ReDim colWidth(1 To columnCount)
```

- **Dynamic Array Initialization**: This line creates an array colWidth that will hold the calculated widths for each column based on the longest text in that column.

Calculate Column Widths

VBA Code 104

```
' CALCULATE COLUMN WIDTHS based on the longest text in each column
    totalTableWidth = 0 ' Initialize total table width
    For colIndex = 1 To columnCount
    maxTextLength = 0
    ' Loop through the rows to find the longest
    'text in the current column
    For x = 1 To lr - 1
    cellText = CStr(mysheet.Cells(x, colIndex).Value)
    If Len(cellText) > maxTextLength Then
    maxTextLength = Len(cellText)
    End If
    Next x
    ' Set column width based on the longest text
```

```
colWidth(colIndex) = maxTextLength * textSize * 1.2
' Adjust this multiplier as needed for extra padding
' Add column width to the total table width
totalTableWidth = totalTableWidth + colWidth(colIndex)
Next colIndex
```

- **Width Calculation**: This loop iterates through each column to calculate the maximum text length in that column. The width for each column is set based on this length, with a multiplier for extra padding.
- **Total Table Width**: It accumulates the total width of all columns to center the table later.

Total Height Calculation

VBA Code 105

```
' CALCULATE THE TOTAL height of the table
    totalTableHeight = (lr - 1) * cellHeight
```

• • • •

- **Height Calculation**: This line calculates the total height of the table based on the number of rows and the height of each cell.

Centering the Table

VBA Code 106

```
' CALCULATE STARTING X and Y positions to center the table
    Dim tableStartX As Double
    Dim tableStartY As Double
    tableStartX = (totalWidth - totalTableWidth) / 2
    ' Center horizontally
    tableStartY = (totalHeight + totalTableHeight) / 2
    ' Center vertically (positive because Y decreases
    'downward in MicroStation)
    ' Initialize cumulativeX with tableStartX to
```

'begin positioning columns from there
cumulativeX = tableStartX

- **Center Position Calculation**: This section calculates where to start drawing the table to center it in the defined area. It computes starting X and Y positions based on the total available space and the dimensions of the table.

Drawing the Table

VBA Code 107

```
' COLUMN LOOP FOR ALL columns in the sheet
    On Error Resume Next ' Ignore any errors that may occur
    For colIndex = 1 To columnCount
    ' Row loop
    For x = 1 To lr - 1
    ' Get cell value as text
    cellText = CStr(mysheet.Cells(x, colIndex).Value)
    ' If it's the first row (title row),
    'convert text to uppercase
    If x = 1 Then
    cellText = UCase(cellText) ' Convert to uppercase
    End If
    ' Create text element
    Set mytext = Application.CreateTextElement1 _
    (Nothing, cellText, Point3dFromXYZ _
    (cumulativeX + paddingLeft, tableStartY - (x - 1) * _
    cellHeight - (cellHeight / 2) + paddingVertical, 0), rotmatrix)
    ActiveModelReference.AddElement mytext
    ' Create table borders (box for each cell)
    ' Top line
    LineSt.x = cumulativeX
    LineSt.y = tableStartY - (x - 1) * cellHeight
    LineSt.z = 0
    LineEn.x = cumulativeX + colWidth(colIndex)
    LineEn.y = tableStartY - (x - 1) * cellHeight
    LineEn.z = 0
    Set myline = _
    Application.CreateLineElement2(Nothing, LineSt, LineEn)
    ActiveModelReference.AddElement myline
```

```
' Bottom line
LineSt.x = cumulativeX
LineSt.y = tableStartY - x * cellHeight
LineSt.z = 0
LineEn.x = cumulativeX + colWidth(colIndex)
LineEn.y = tableStartY - x * cellHeight
LineEn.z = 0
Set myline = _
Application.CreateLineElement2(Nothing, LineSt, LineEn)
ActiveModelReference.AddElement myline
' Right line
LineSt.x = cumulativeX + colWidth(colIndex)
LineSt.y = tableStartY - (x - 1) * cellHeight
LineSt.z = 0
LineEn.x = cumulativeX + colWidth(colIndex)
LineEn.y = tableStartY - x * cellHeight
LineEn.z = 0
Set myline = _
Application.CreateLineElement2(Nothing, LineSt, LineEn)
ActiveModelReference.AddElement myline
' Left line
LineSt.x = cumulativeX
LineSt.y = tableStartY - (x – 1)
```

While you already have a substantial VBA code for this process, the main sections are:

- **Variable Declarations**: To hold data types and references to Excel and MicroStation elements.
- **Excel Connection**: To connect and navigate the Excel application and worksheet.
- **Data Processing**: To iterate through the cells in Excel, calculating dimensions for the table.
- **Element Creation**: To add text and lines in MicroStation to create the visual representation of the table.

Benefits of This Approach

1. **Efficiency**: Automating the transfer of data reduces the time spent on

manual entry and minimizes errors.
2. **Consistency**: Ensures that the data presented in the drawing matches the data in the Excel sheet, maintaining consistency across documents.
3. **Flexibility**: Easily update data in Excel, rerun the macro, and have the table in MicroStation reflect the changes without manual adjustments.
4. **Professional Presentation**: Creates clean and organized tables that enhance the visual quality of your drawings.

Conclusion

USING VBA TO TRANSFER data from Excel to MicroStation allows for a seamless integration of tabular data into your design workflow. By automating this process, you can save time, reduce errors, and maintain consistency across your projects. This approach is particularly useful for engineers, architects, and designers who often work with large datasets and require clear presentation in their drawings.

22.
Exporting and Importing Data Between MicroStation and Notepad Using .txt Files

In this chapter, we will explore how to **export** data from MicroStation into a .txt file, and how to **import** data from a .txt file back into MicroStation. These techniques are useful for saving coordinate data, element properties, or logs and using them in other systems or applications.

1. Exporting MicroStation Data to a .txt File

WE WILL COVER DIFFERENT scenarios for exporting data from MicroStation, including exporting **text elements**, **element properties**, and **coordinates**.

1.1. Exporting Text Elements and Properties to a .txt File

THIS EXAMPLE SHOWS how to scan for text elements in a MicroStation model, retrieve their **element ID**, **text content**, and **level name**, and export this data to a .txt file.

VBA Code 108

```
SUB EXPORTMICROSTATIONTEXTTOTXT()
    Dim filePath As String
    Dim fileNum As Integer
    Dim mytext As TextElement
    Dim MyScanCriteria As ElementScanCriteria
    Dim myScanEnum As ElementEnumerator
    Dim oMicroStation As Object
    ' Specify the file path for the .txt file
    filePath = "C:\path\to\your\exported_data.txt"
    ' Open the text file for writing
    fileNum = FreeFile
    Open filePath For Output As fileNum
    ' Create a connection to MicroStation
    On Error Resume Next
```

```
Set oMicroStation = GetObject(, "MicroStationDGN.Application")
If oMicroStation Is Nothing Then
MsgBox "MicroStation is not running."
Exit Sub
End If
On Error GoTo 0
' Set up scan criteria to include only text elements
Set MyScanCriteria = New ElementScanCriteria
MyScanCriteria.ExcludeAllTypes
MyScanCriteria.IncludeType msdElementTypeText
' Scan for text elements in the active model
Set myScanEnum = _
oMicroStation.ActiveModelReference.Scan(MyScanCriteria)
' Loop through each text element found and write
'it to the .txt file
Do While myScanEnum.MoveNext
Set mytext = myScanEnum.Current
' Write element ID and text content to the .txt file
Print #fileNum, "Element ID: " & mytext.ID64
Print #fileNum, "Text Content: " & mytext.Text
Print #fileNum, "———————————————————."
Loop
' Close the text file
Close fileNum
' Notify the user
MsgBox "Data successfully exported to " & filePath
End Sub
```

This subroutine exports the **ID**, **text content**, and **level** of text elements from MicroStation into a text file.

1.2. Exporting Element Coordinates to a .txt File

SOMETIMES, YOU NEED to export the coordinates of elements, such as **LineString** elements, to a .txt file. This example shows how to do that:

VBA Code 109

```
SUB SAVEELEMENTCOORDINATESTOTEXTFILE()
    Dim element As Element
    Dim elementEnumerator As ElementEnumerator
```

```vba
Dim point As Point3d
Dim i As Long
Dim filePath As String
Dim fileNumber As Integer
Dim lineData As String
Dim vertex As Point3d
Dim filter As ElementScanCriteria
' File path to save coordinates (Desktop)
filePath = _
Environ("USERPROFILE") & "\Desktop\element_coordinates.txt"
' Open the file for writing
fileNumber = FreeFile
Open filePath For Output As fileNumber
' Create and configure the scan criteria
Set filter = New ElementScanCriteria
filter.ExcludeAllTypes ' Exclude all element types first
filter.IncludeType msdElementTypeLineString
' Include only LineString elements
' Scan the model for elements that match the criteria
Set elementEnumerator = ActiveModelReference.Scan(filter)
' Loop through each element found by the scan
While elementEnumerator.MoveNext
Set element = elementEnumerator.Current
' Process LineString elements
If IsLineStringElement(element) Then
Dim lineString As LineElement
Set lineString = element.AsLineElement
' Loop through all vertices of the line
'string and write them
For i = 1 To lineString.VerticesCount
vertex = lineString.vertex(i)
lineData = _
vertex.X & "," & vertex.Y & "," & vertex.Z
Print #fileNumber, lineData
Next i
End If
Wend
' Close the file
Close fileNumber
' Inform the user
MsgBox "Element coordinates saved to: " & filePath
' Open the file automatically in the default text editor
```

```
Shell "notepad.exe " & filePath, vbNormalFocus
End Sub
' Function to check if an element is a LineStringElement
Function IsLineStringElement(elm As Element) As Boolean
On Error Resume Next
IsLineStringElement = Not elm.AsLineElement Is Nothing
On Error GoTo 0
End Function
```

This subroutine exports the **coordinates** of vertices from LineString elements into a text file, where each vertex is saved as an X,Y,Z coordinate on a new line.

2. Importing Data from a .txt File into MicroStation

YOU CAN ALSO IMPORT data from a .txt file into MicroStation, such as loading coordinate data to create elements directly in the MicroStation model.

2.1. Importing Coordinates from a .txt File

IN THIS EXAMPLE, YOU will import coordinates from a .txt file and use them to place a **LineString** in MicroStation.

VBA Code 110

```
SUB TXT_TO_STRING()
    Dim startPoint As Point3d
    Dim point As Point3d
    Dim filePath As String
    Dim fileNumber As Integer
    Dim lineData As String
    Dim coordinates() As String
    Dim i As Long
    ' File path to your coordinates.txt (adjust accordingly)
    filePath = Environ("USERPROFILE") & "\Desktop\coordinates.txt"
    ' Open the file for reading
    fileNumber = FreeFile
    Open filePath For Input As fileNumber
    ' Start the line string command
    CadInputQueue.SendKeyin "place lstring point"
```

```
' Read each line from the file
i = 0
Do While Not EOF(fileNumber)
Line Input #fileNumber, lineData
coordinates = Split(lineData, ",")
' Parse coordinates
point.X = CDbl(coordinates(0))
point.Y = CDbl(coordinates(1))
point.Z = CDbl(coordinates(2))
' Send the data point to the current command
CadInputQueue.SendDataPoint point, 1
i = i + 1
Loop
' Close the file
Close fileNumber
' Send a reset to the current command
CadInputQueue.SendReset
' End the command
CommandState.StartDefaultCommand
End Sub
```

This subroutine reads coordinates from a text file, interprets them as X, Y, Z points, and uses them to place a **LineString** in MicroStation.

3. Benefits of Using .txt Files for Data Transfer

- **Simplicity**: Text files are easy to create, read, and edit, and they don't require complex data structures or software to open and modify.
- **Lightweight**: Text files are smaller compared to Excel or databases, making them faster to process and easier to share.
- **Compatibility**: .txt files are universally readable by all operating systems and applications, providing great flexibility for data exchange.

Chapter Summary

IN THIS CHAPTER, WE covered how to:

1. **Export text elements** and **element coordinates** from MicroStation to a .txt file.

2. **Import coordinate data** from a .txt file into MicroStation to create elements.
3. **Handle and automate** processes, such as automatically opening exported files in Notepad.

Using .txt files for exporting and importing data between MicroStation and external applications provides a simple and effective way to handle data transfer.

23. VBA Calculations in MicroStation

VBA (Visual Basic for Applications) in MicroStation allows users to perform advanced calculations on design elements such as circles, lines, shapes, and solids. These calculations can include measuring distances, calculating areas and volumes, determining angles between elements, and computing statistical properties of design geometry. This chapter will demonstrate how to use VBA to automate common geometric calculations in MicroStation.

1. Distance Calculation Between Two Circles

TO MEASURE THE DISTANCE between the centers of two circles in MicroStation, you can fetch their centroids and use the **Pythagorean theorem** to calculate the distance.

Code Example: Calculating Distance Between Two Circles

VBA Code 111

```
FUNCTION CALCULATEDISTANCEFROMELEMENTS _
    (el1 As Element, el2 As Element) As Double
    Dim point1 As Point3d
    Dim point2 As Point3d
    ' Get the center of the first circle element
    point1 = el1.AsEllipseElement.Centroid
    ' Get the center of the second circle element
    point2 = el2.AsEllipseElement.Centroid
    ' Calculate the distance using the Pythagorean theorem
    CalculateDistanceFromElements = _
    Sqr((point2.x - point1.x) ^ 2 + _
    (point2.y - point1.y) ^ 2 + (point2.Z - point1.Z) ^ 2)
    ' Print the result in the Immediate window
    ' (for debugging purposes)
    Debug.Print "Distance between circles: " & _
    CalculateDistanceFromElements
```

```
End Function
Sub CalculateDistanceBetweenCircles()
Dim el1 As element
Dim el2 As element
Dim distance As Double
Dim circleID1 As Long
Dim circleID2 As Long
' Set the IDs of the two circle elements
circleID1 = 477375 ' Replace with the first circle's ID
circleID2 = 477376 ' Replace with the second circle's ID
' Fetch the first circle by ID
Set el1 = ActiveModelReference.GetElementByID _
(DLongFromString(CStr(circleID1)))
' Fetch the second circle by ID
Set el2 = ActiveModelReference.GetElementByID _
(DLongFromString(CStr(circleID2)))
' Calculate the distance between the two circles
If Not el1 Is Nothing And Not el2 Is Nothing Then
distance = CalculateDistanceFromElements(el1, el2)
MsgBox "The distance between the two circles is: " _
& distance
Else
MsgBox _
"One or both of the circle elements could not be found."
End If
End Sub
```

Explanation:

- The function CalculateDistanceFromElements retrieves the centroids of the two circle elements and calculates the distance between them using the **Pythagorean theorem**.
- The CalculateDistanceBetweenCircles subroutine fetches the circle elements by their IDs and invokes the distance calculation function.

2. Calculating Area of a Shape

VBA CAN AUTOMATE THE process of calculating the area of a 2D shape or the volume of a 3D solid. This allows designers to quickly retrieve geometric properties of elements.

Code Example: Calculating Area

VBA Code 112

```
SUB CALCULATESHAPEAREA()
    Dim shapeElement As element
    Dim shapeArea As Double
    Dim shapeID As Long
    ' Set the ID for the shape element
    shapeID = 477378 ' Replace with the shape element's ID
    ' Fetch the shape element by ID
    Set shapeElement = _
    ActiveModelReference.GetElementByID(DLongFromString(CStr(shapeID)))
    ' Calculate the area of the shape element
    If Not shapeElement Is Nothing Then
    shapeArea = CalculateShapeAreaFromElement(shapeElement)
    ' Ensure the function name matches
    If shapeArea <> -1 Then
    ' Check if the area calculation was successful
    MsgBox "The area of the shape element is: " & shapeArea
    Else
    MsgBox "Area calculation failed."
    End If
    Else
    MsgBox "The shape element could not be found."
    End If
End Sub
```

Explanation:

- The subroutine CalculateShapeArea retrieves the shape by its ID and calls the area calculation function.

3. Calculating the Angle Between Two Line Elements

USING VBA, YOU CAN compute the angle between two **LineElements** based on their vertices using vector mathematics. This is essential when working with elements that need to follow specific angular constraints.

Code Example: Calculating Angle Between Line Elements

VBA Code 113

```
FUNCTION CALCULATEANGLEBETWEENELEMENTS _
    (el1 As Element, el2 As Element) As Double
    Dim vertices1() As Point3d
    Dim vertices2() As Point3d
    Dim vector1 As Point3d
    Dim vector2 As Point3d
    Dim dotProduct As Double
    Dim magnitude1 As Double
    Dim magnitude2 As Double
    ' Check if both elements are LineElements
    If el1.IsLineElement And el2.IsLineElement Then
    ' Get the vertices of both LineElements
    vertices1 = el1.AsLineElement.GetVertices
    vertices2 = el2.AsLineElement.GetVertices
    ' Calculate direction vectors for both lines
    vector1.X = vertices1(1).X - vertices1(0).X
    vector1.Y = vertices1(1).Y - vertices1(0).Y
    vector1.Z = vertices1(1).Z - vertices1(0).Z
    vector2.X = vertices2(1).X - vertices2(0).X
    vector2.Y = vertices2(1).Y - vertices2(0).Y
    vector2.Z = vertices2(1).Z - vertices2(0).Z
    ' Calculate the dot product of the two vectors
    dotProduct = (vector1.X * vector2.X + _
    vector1.Y * vector2.Y + vector1.Z * vector2.Z)
    ' Calculate the magnitude of each vector
    magnitude1 = _
    Sqr(vector1.X ^ 2 + vector1.Y ^ 2 + vector1.Z ^ 2)
    magnitude2 = _
    Sqr(vector2.X ^ 2 + vector2.Y ^ 2 + vector2.Z ^ 2)
```

```
' Calculate the angle between the two
'vectors using the dot product formula
If magnitude1 > 0 And magnitude2 > 0 Then
CalculateAngleBetweenElements = _
Atn(dotProduct / (magnitude1 * magnitude2))
CalculateAngleBetweenElements = _
CalculateAngleBetweenElements * (180 / 3.14159265358979)
' Convert to degrees
Else
MsgBox _
"One or both of the vectors have zero magnitude."
CalculateAngleBetweenElements = -1
End If
Else
MsgBox "One or both elements are not LineElements."
CalculateAngleBetweenElements = -1 ' Return -1 to indicate error
End If
End Function
Sub CalculateAngleBetweenTwoElements()
Dim el1 As element
Dim el2 As element
Dim angle As Double
Dim elementID1 As Long
Dim elementID2 As Long
' Set the IDs of the two elements
elementID1 = 468401 ' Replace with the first element's ID
elementID2 = 468402 ' Replace with the second element's ID
' Fetch the first element by ID
Set el1 = ActiveModelReference.GetElementByID _
(DLongFromString(CStr(elementID1)))
' Fetch the second element by ID
Set el2 = ActiveModelReference.GetElementByID _
(DLongFromString(CStr(elementID2)))
' Calculate the angle between the two elements
If Not el1 Is Nothing And Not el2 Is Nothing Then
angle = CalculateAngleBetweenElements(el1, el2)
If angle <> -1 Then
MsgBox "The angle between the two elements is: " _
& angle & " degrees."
End If
Else
MsgBox "One or both elements could not be found."
```

End If
End Sub

Explanation:

- **Direction vectors** are computed by subtracting the coordinates of the start points from the end points of each line element.
- The **dot product** and the **magnitude** of each vector are calculated, and the angle between the two lines is derived from these values.

4. Calculating the Average Distance of Vertices in a LineString

YOU CAN CALCULATE THE average distance of the vertices in a **LineString** element by fetching its points and averaging their distances from the origin.

Code Example: Calculating Average Distance in a LineString

VBA Code 114

```
FUNCTION CALCULATEAVERAGEFROMLINESTRING(el As Element) As Double
    Dim vertices() As Point3d
    Dim sum As Double
    Dim i As Integer
    ' Check if the element is a LineStringElement
    If TypeOf el Is LineElement Then
    ' Get the vertices of the LineString element
    vertices = el.AsLineElement.GetVertices
    ' Check that there are vertices
    If UBound(vertices) >= 0 Then
    ' Sum the distances of points in the array
    For i = LBound(vertices) To UBound(vertices)
    sum = sum + Sqr(vertices(i).X ^ 2 + _
    vertices(i).Y ^ 2 + vertices(i).Z ^ 2)
    Next i
    ' Return the average distance
    CalculateAverageFromLineString = sum / _
```

```
        (UBound(vertices) - LBound(vertices) + 1)
    Else
        MsgBox "The element does not have any vertices."
        CalculateAverageFromLineString = -1
        ' Return -1 to indicate error
    End If
Else
    MsgBox "The element is not a LineStringElement."
    CalculateAverageFromLineString = -1
    ' Return -1 to indicate error
End If
End Function
Sub CalculateAverageDistanceFromLineString()
    Dim el As element
    Dim averageDistance As Double
    Dim elementID As Long
    ' Set the ID of the LineString element
    elementID = 468403
    ' Replace with the actual LineString element's ID
    ' Fetch the LineString element by ID
    Set el = ActiveModelReference.GetElementByID _
    (DLongFromString(CStr(elementID)))
    ' Calculate the average distance from the LineString vertices
    If Not el Is Nothing Then
        averageDistance = CalculateAverageFromLineString(el)
        If averageDistance <> -1 Then
            MsgBox _
            "The average distance of the points is: " & averageDistance
        End If
    Else
        MsgBox "The LineString element could not be found."
    End If
End Sub
```

Explanation:

- The function CalculateAverageFromLineString calculates the average distance of all vertices in a **LineString** from the origin.

Chapter Summary: VBA Calculations in

MicroStation

IN THIS CHAPTER, WE explored how VBA can be used in MicroStation to perform several essential calculations. These include:

1. **Calculating the distance** between two circle elements.
2. **Calculating the area** of shape elements.
3. **Determining the angle** between two-line elements using vector mathematics.
4. **Calculating the average distance** of vertices in a **LineString** element.

By automating these calculations using VBA, designers can streamline workflows, reduce manual computation errors, and ensure efficient handling of complex geometry in MicroStation. The examples demonstrated in this chapter can be expanded upon to suit more advanced needs, such as incorporating these calculations into larger automation scripts for entire projects.

24.
Placing Cells in 2D and 3D, and Automating with Excel Data in MicroStation

In this chapter, we explore how to place cells in both 2D and 3D environments using VBA in MicroStation. Additionally, we will demonstrate how to automate the placement of multiple cells using data from an Excel sheet, enhancing productivity and minimizing errors in large-scale projects.

Placing Cells in a 2D

CELLS IN MICROSTATION are reusable components, such as symbols, objects, or details, that can be inserted into different designs. In a 2D environment, you can place cells at specific coordinates using VBA. This method involves selecting a cell from a predefined cell library and placing it at a desired point within the design.

Example of Placing a Cell in a 2D Model:

VBA Code 115

```
SUB MAIN_PROCEDURE_2D()
    ' Calling the PlaceCell2D subroutine with arguments
    PlaceCell2D "2d_cell_name", 10, 20, 1, 45
    End Sub
    Sub PlaceCell2D(cellName As String, x As Double, _
    y As Double, Pscale As Double, angle As Double)
    Dim oCell As CellElement
    Dim cellOrigin As Point3d
    Dim RotMatrix As Matrix3d
    Dim given_scale As Point3d
    Dim radians As Double
    ' Set the scale
    given_scale = Point3dFromXYZ(Pscale, Pscale, Pscale)
    ' Uniform scaling
```

```
' Convert angle from degrees to radians
radians = angle * (3.14159265358979 / 180)
' Convert degrees to radians
' Create the rotation matrix around the Z-axis
RotMatrix = Matrix3dFromAxisAndRotationAngle(2, radians)
' 2 for Z-axis
' Set the origin at the 2D coordinates
cellOrigin = Point3dFromXYZ(x, y, 0)
' Create and place the cell at the specified origin
Set oCell = CreateCellElement2 _
(cellName, cellOrigin, given_scale, True, RotMatrix)
ActiveModelReference.AddElement oCell
End Sub
```

Explanation:

- **cellName**: Name of the cell to place, which must exist in the active cell library.
- **x and y**: The 2D coordinates where the cell will be placed.
- **CreateCellElement2**: This function creates a cell element based on its name and origin. The **Matrix3dIdentity** keeps the cell in its default orientation.
- **ActiveModelReference.AddElement**: Adds the created cell to the active 2D drawing.

Best Practices for Placing Cells in 2D:

- **Precise Origin Selection**: Ensure the origin of the cell is carefully selected to control the cell's final position in the drawing.
- **Scaling and Rotation**: Apply transformations to resize or rotate the cell if needed, such as scaling by a factor or rotating around a specific axis.
- **Layer Management**: Place cells on the correct layer (level) for better element management in complex drawings.

Additional Notes:

- **Check for Cell Existence**: Before placing a cell, verify that it exists in the active cell library to avoid runtime errors.

Placing Cells in a 3D Model

CELLS CAN ALSO BE PLACED in a 3D environment, allowing you to control their position along the Z-axis in addition to X and Y coordinates. This method is commonly used for placing structural elements, mechanical components, and architectural details in 3D space.

Example of Placing a Cell in a 3D Model:

VBA Code 116

```
SUB MAIN_PROCEDURE_3D()
    ' Calling the PlaceCell3D subroutine with arguments
    PlaceCell3D "3d_cell_name", 10, 20, 5, 1, 30
    End Sub
    Sub PlaceCell3D(cellName As String, x As Double, _
    y As Double, z As Double, Pscale As Double, angle As Double)
    Dim oCell As CellElement
    Dim cellOrigin As Point3d
    Dim RotMatrix As Matrix3d
    Dim given_scale As Point3d
    Dim radians As Double
    ' Set the scale
    given_scale = Point3dFromXYZ(Pscale, Pscale, Pscale)
    ' Uniform scaling
    ' Convert angle from degrees to radians
    radians = angle * (3.14159265358979 / 180)
    ' Convert degrees to radians
    ' Create the rotation matrix around the Z-axis
    RotMatrix = Matrix3dFromAxisAndRotationAngle(2, radians)
    ' 3 for Z-axis
    ' Set the origin at the 3D coordinates
    cellOrigin = Point3dFromXYZ(x, y, z)
    ' Include Z-coordinate
```

```
' Create and place the cell at the specified origin
Set oCell = CreateCellElement2 _
(cellName, cellOrigin, given_scale, True, RotMatrix)
ActiveModelReference.AddElement oCell
End Sub
```

Explanation:

- **x, y, and z**: These are the 3D coordinates where the cell will be placed.
- **CreateCellElement2**: Creates the cell at the specified 3D origin, with default orientation maintained by the **Matrix3dIdentity**.
- **ActiveModelReference.AddElement**: Adds the 3D cell into the active 3D model.

Best Practices for 3D Cells:

- **Coordinate Validation**: Validate the X, Y, and Z coordinates to ensure precise placement within the 3D environment.
- **3D Rotation and Scaling**: Consider rotating or scaling the cell to fit the 3D environment. You can apply transformation matrices to achieve this.
- **Z-Level Consideration**: Be mindful of the Z-axis, especially when placing elements at ground level, elevated positions, or below the surface.
- **Collision Detection**: Ensure cells are spaced appropriately to avoid overlaps or visual errors in complex 3D models.

Advanced 3D Cell Placement:

- **Rotation Matrices**: You can rotate cells by creating a custom rotation matrix before placement:

VBA Code 117

DIM ROTATIONMATRIX As Matrix3d

```
rotationMatrix = _
RotMatrix3dFromAxisAndRotationAngle _
(Point3dFromXYZ(0, 0, 1), Radians(45))
' Rotate 45 degrees around Z-axis
Set oCell = CreateCellElement2 _
(cellName, cellOrigin, rotationMatrix, True)
```

- **Scaling Matrices**: You can scale a cell by applying a scaling transformation matrix:

VBA Code 118

```
DIM SCALEMATRIX AS Matrix3d
    scaleMatrix = Matrix3dFromScaleFactors(2.0, 2.0, 2.0)
    ' Double the size in all directions
    Set oCell = CreateCellElement2 _
    (cellName, cellOrigin, scaleMatrix, True)
```

Placing a Cell Using UserForm.

NOW WE KNOW HOW TO place cell using a code, lets create a form to place a Cell. This section shows how to read from an existing cell library and place cells using a UserForm. This part involves three components: UserForm, Module, and Class Module. We need to design a UserForm with Cell Name, Coordinates and Place Cell button.

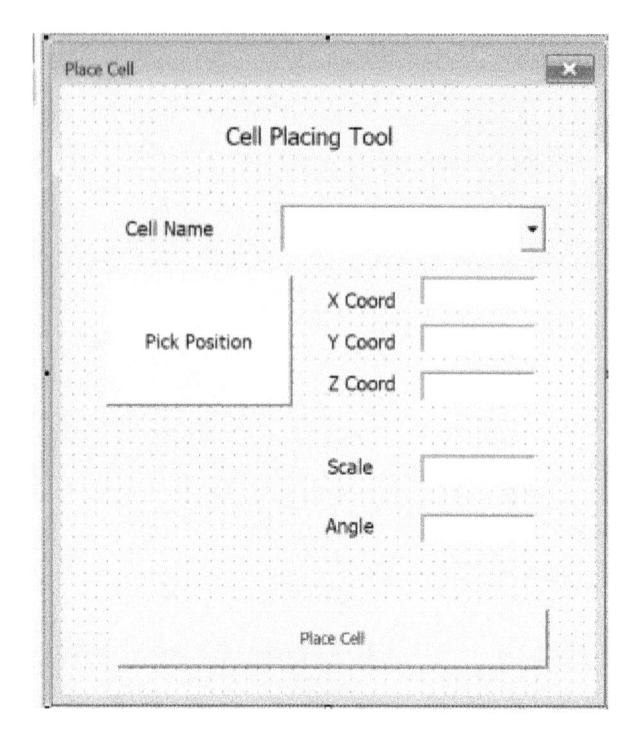

Figure 21 - Cell Placing Form

UserForm Design:

CREATE A USERFORM NAMED frmPlaceCell with the following elements:

- **Pick Position Button**: btn_pick_position to pick X, Y, Z positions in the model.
- **Place Cell Button**: btn_place_cell to place the cell.
- **Cell Name ComboBox**: cmb_cell_list to list the cells from the cell library.
- **X, Y, Z Coordinate Text Boxes**: txt_x, txt_y, txt_z to enter respective coordinates.
- **Scale and Angle Text Boxes**: to enter scale and rotation angle.
- **Labels**: to guide the user on each input.

Add Module and call it "placing_cells_with_form" and add following code

Showing the User form

VBA Code 119

```
SUB SHOWCELLPLACEMENTFORM()
    frmPlaceCell.Show modeless
End Sub
```

This subroutine opens the cell placement form (frmPlaceCell) in a modeless way, allowing users to interact with other windows while keeping the form open.

Placing a 3D Cell

THE CODE BELOW REQUIRES 6 inputs to place the cells as shown.

VBA Code 120

```
SUB PLACECELL3D_WITH_form(cellName As String, x As Double, _
    y As Double, z As Double, Pscale As Double, angle As Double)
    Dim oCell As CellElement
    Dim cellOrigin As Point3d
    Dim RotMatrix As Matrix3d
    Dim given_scale As Point3d
    Dim radians As Double
    ' Set the scale
    given_scale = Point3dFromXYZ(Pscale, Pscale, Pscale)
    ' Uniform scaling
    ' Convert angle from degrees to radians
    radians = angle * (3.14159265358979 / 180)
    ' Create the rotation matrix around the Z-axis
    RotMatrix = Matrix3dFromAxisAndRotationAngle(2, radians)
    ' 2 for Z-axis
    ' Set the origin at the 3D coordinates
    cellOrigin = Point3dFromXYZ(x, y, z)
    ' Create and place the cell at the specified origin
    Set oCell = CreateCellElement2 _
        (cellName, cellOrigin, given_scale, True, RotMatrix)
```

```
ActiveModelReference.AddElement oCell
End Sub
```

This subroutine takes the cell name, coordinates, scale, and rotation angle to create and place a 3D cell in the model. It calculates the scale and rotation matrix based on the provided parameters and then creates the cell element in the specified position.

Getting Coordinates from User Input

THE CODE BELOW TAKE the names of the text box from the form and returns x,y,z coordinates.

VBA Code 121

```
SUB GET_COORDINATES(ByRef txtX As TextBox, _
    ByRef txtY As TextBox, ByRef txtZ As TextBox)
    Dim MyQue As CadInputQueue
    Dim SelPt As Point3d
    ' Clear old data
    txtX.Text = ""
    txtY.Text = ""
    txtZ.Text = ""
    On Error GoTo errhnd
    Set MyQue = Application.CadInputQueue
    Do
    Dim MyMsg As CadInputMessage
    Set MyMsg = MyQue.GetInput
    Select Case MyMsg.InputType
    Case msdCadInputTypeDataPoint
    SelPt = MyMsg.point
    txtX.Text = SelPt.x
    txtY.Text = SelPt.y
    txtZ.Text = SelPt.z
    Exit Do
    Case Else
    Exit Do
    End Select
    Loop
    Exit Sub
    errhnd:
```

INTRODUCTION TO MICROSTATION VBA

Err.Clear
End Sub

This subroutine retrieves coordinates from the user's mouse clicks and populates the text boxes (txtX, txtY, and txtZ) with the selected point's coordinates. It uses a loop to wait for user input and handles errors gracefully.

Get Cell location using Modal Dialog Events

CELL LOCATION WITHIN your directory can be called and opened by creating a class module. Create a new class module and call it "cell_location"

Figure 22 - Class Module

ADD THE FOLLOWING CODE.

VBA Code 122

```
IMPLEMENTS IMODALDIALOGEVENTS
    Private Sub IModalDialogEvents_OnDialogClosed _
    (ByVal DialogBoxName As String, ByVal DialogResult _
    As MsdDialogBoxResult)
    End Sub
    Private Sub IModalDialogEvents_OnDialogOpened _
    (ByVal DialogBoxName As String, DialogResult As MsdDialogBoxResult)
    If DialogBoxName = "Attach Cell Library" Then
    CadInputQueue.SendCommand          "MDL          COMMAND
MGDSHOOK,fileList_setDirectoryCmd C:\Cells\"
    CadInputQueue.SendCommand          "MDL          COMMAND
MGDSHOOK,fileList_setFileNameCmd MyCellLibrary.dgn"
    DialogResult = msdDialogBoxResultOK
    End If
    End Sub
```

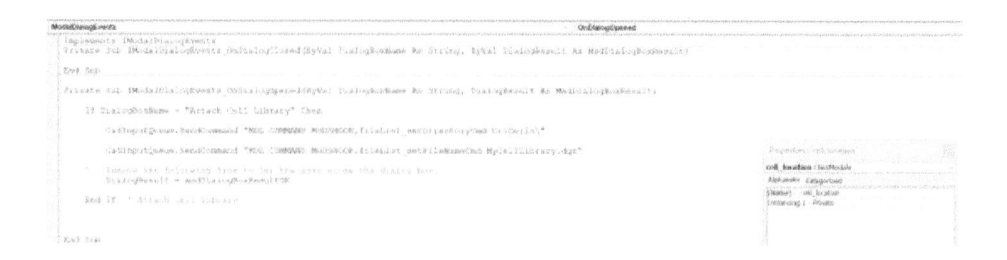

THIS SECTION MANAGES events related to modal dialogs, specifically when a dialog for attaching a cell library is opened. It sets the directory and filename for the library to be used.

UserForm Events

ADD THE REST OF THE code to the UserForm buy double clicking the form itself.

VBA Code 123

```
PRIVATE SUB USERFORM_Initialize()
    Dim modalHandler As New cell_location
    AddModalDialogEventsHandler modalHandler
    CadInputQueue.SendCommand "ATTACH LIBRARY"
    RemoveModalDialogEventsHandler modalHandler
    CommandState.StartDefaultCommand
    ' Now enumerate the cells in the attached library
    Dim cellEnumerator As CellInformationEnumerator
    Dim cellInfo As CellInformation
    ' Assuming you've attached the library successfully,
    'we get the cell enumerator
    Set cellEnumerator = _
    Application.GetCellInformationEnumerator(True, True)
    ' Clear the ComboBox before populating
    cmb_cell_list.Clear
    ' Loop through the cells in the library and
    'add them to the ComboBox
    Do While cellEnumerator.MoveNext
    Set cellInfo = cellEnumerator.Current
    cmb_cell_list.AddItem cellInfo.Name
```

```
Loop
txt_x.value = 0
txt_y.value = 0
txt_z.value = 0
txt_scale.value = 1
txt_angle.value = 0
End Sub
Private Sub cmb_pick_position_Click()
' Call the function to get coordinates
placing_cells_with_form.get_coordinates txt_x, txt_y, txt_z
End Sub
Private Sub cmb_place_cell_Click()
placing_cells_with_form.PlaceCell3D_with_form _
cmb_cell_list, txt_x.value, txt_y.value, _
txt_z.value, txt_scale.value, txt_angle.value
End Sub
```

- **cmb_pick_position_Click**: This event triggers when the user clicks the button to pick a position, calling the get_coordinates function to update the text boxes with the selected point.
- **cmb_place_cell_Click**: This event places the selected cell at the specified coordinates, scale, and angle when the button is clicked.
- **UserForm_Initialize**: This initializes the form by attaching the cell library and populating the combo box with available cell names from the library.

Summary

THIS CHAPTER COVERS the process of placing 3D cells interactively using a form, detailing the key subroutines and their roles. Each component works together to provide a smooth user experience for selecting and placing cells in the drawing environment.

Placing Multiple Cells from Excel Data in MicroStation

IN MANY DESIGN PROJECTS, you may need to place multiple cells at different locations based on predefined data. In some scenarios, you may want to apply rotation or scaling to the cells based on data in the Excel sheet (e.g., rotation angles or scale factors).

By using Excel to manage cell names and their corresponding coordinates, you can automate the placement of multiple cells in MicroStation, streamlining your workflow.

Example Structure of Excel Data:

Cell Name	X Coordinate	Y Coordinate	Z Coordinate	Scale	Rotation
Cell_1	0	0	0	1	0
Cell_2	0	5	5	1	45
Cell_3	0	10	10	1	90

THIS CODE SHOULD PLACE multiple cells in the model, Excel file must be opened and filled in as shown on the table above.

Example of Placing Cells with Rotation and Scaling:

VBA Code 124

```vb
SUB PLACECELLSFROMEXCEL()
    Dim xlApp As Excel.Application
    Dim wb As Excel.Workbook
    Dim ws As Excel.Worksheet
    Dim row As Long
    Dim lastRow As Long
    Dim cellName As String
    Dim x As Double
    Dim y As Double
    Dim z As Double
    Dim angle As Double
    Dim Pscale As Double
    ' Get the active Excel application and workbook
    Set xlApp = GetObject(, "Excel.Application")
    Set wb = xlApp.ActiveWorkbook
    Set ws = wb.ActiveSheet ' Use the active sheet directly
    ' Find the last row with data in the active sheet
    lastRow = ws.Cells(ws.Rows.Count, 1).End(xlUp).row
    ' Loop through each row to get the cell name and coordinates
    For row = 2 To lastRow
    ' Read the cell name and coordinates from Excel
    cellName = ws.Cells(row, 1).value
    x = ws.Cells(row, 2).value
    y = ws.Cells(row, 3).value
    z = ws.Cells(row, 4).value ' Z coordinate
    Pscale = ws.Cells(row, 5).value
    ' Angle, if you want to use it
    angle = ws.Cells(row, 6).value ' Set the scale as needed
    ' Call PlaceCell3D to place the cell
    Call PlaceCell3D(cellName, x, y, z, Pscale, angle)
    Next row
    ' Notify the user that the cells have been placed
    MsgBox "Cells have been placed based on Excel data."
    End Sub
    Sub PlaceCell3D(cellName As String, x As Double, _
    y As Double, z As Double, Pscale As Double, angle As Double)
```

```
Dim oCell As CellElement
Dim cellOrigin As Point3d
Dim RotMatrix As Matrix3d
Dim given_scale As Point3d
Dim radians As Double
' Set the scale
given_scale = Point3dFromXYZ(Pscale, Pscale, Pscale)
' Uniform scaling
' Convert angle from degrees to radians
radians = angle * (3.14159265358979 / 180)
' Convert degrees to radians
' Create the rotation matrix around the Z-axis
RotMatrix = Matrix3dFromAxisAndRotationAngle(2, radians)
' 2 for Z-axis
' Set the origin at the 3D coordinates
cellOrigin = Point3dFromXYZ(x, y, z) ' Include Z-coordinate
' Create and place the cell at the specified origin
Set oCell = CreateCellElement2 _
(cellName, cellOrigin, given_scale, True, RotMatrix)
ActiveModelReference.AddElement oCell
End Sub
```

Summary

IN THIS CHAPTER, WE explored how to place cells in both **2D** and **3D** environments using VBA and demonstrated how to automate the placement of multiple cells using Excel data. Key takeaways include:

1. **2D Cell Placement**: Cells can be placed at any X and Y coordinates, and their origin, scaling, and rotation can be carefully controlled.
2. **3D Cell Placement**: Placing cells in 3D involves managing their positioning along the Z-axis and handling rotation and scaling transformations.
3. **Automating Cell Placement with Excel**: Automating the process by reading cell names and coordinates from Excel significantly improves productivity, especially for large-scale projects.
4. **Advanced Transformations**: You can apply both rotation and scaling to each cell for better control over their placement in complex

models.

By leveraging VBA and Excel together, you can streamline repetitive tasks, ensure accuracy, and maintain control over the design elements in both 2D and 3D environments.

25. Recording Macro Tool

Using MicroStation's Macro Recording Tool

- MicroStation features a powerful built-in macro recorder that enables users to capture repetitive tasks and convert them into VBA code automatically. This tool is invaluable for quickly generating scripts for common tasks without extensive programming knowledge.

Steps to Record a Macro

1. **Accessing the Macro Recorder**: Navigate to Utilities > Macro > Record. This opens the macro recording interface.
2. **Performing Tasks**: Execute the tasks you wish to automate, such as drawing elements, placing cells, or modifying properties. The macro recorder tracks your actions in real-time.
3. **Stopping the Recording**: Once you have completed the desired tasks, stop the recording. You will be prompted to save the macro with a specific name.
4. **Reviewing the Code**: Open the VBA editor to review the generated code. The macro may contain useful snippets of code that can be edited or refined to suit more complex automation needs.

Best Practices for Macro Recording

- **Keep It Simple**: Use the macro recorder for straightforward tasks. For more intricate processes, refine the generated code manually to enhance functionality and performance.
- **Test Thoroughly**: Run the recorded macro multiple times in various scenarios to ensure it operates as expected. Look for edge cases that might cause the macro to fail.
- **Documentation**: After refining your macro, document its purpose, functionality, and any parameters it requires. This will aid future

users and provide context for your own reference.

Example: Recording a Light Zone Macro

In this example, I will create a macro that visually represents a light zone when it has been switched on, using a shape that can be adjusted based on the light specifications.

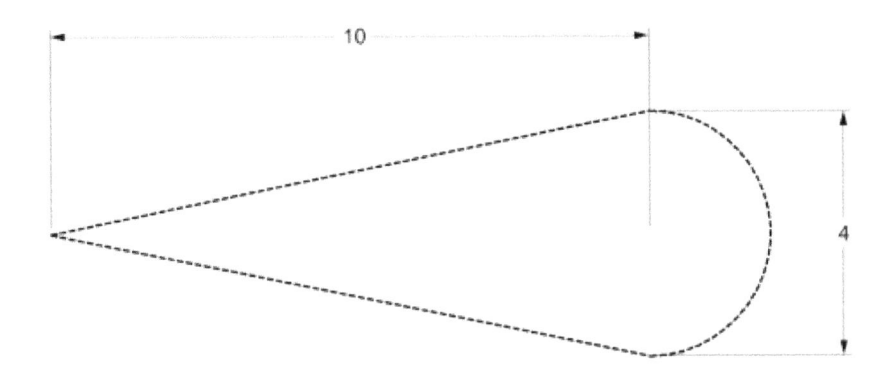

Figure 23 - Light Zone Shape Dimensions

Instructions:

1. STARTING POINT: I will start at coordinates (0,0,0) and aim to create the shape accurately.
   2. Drawing the Shape:

- First, I will create the top line.
- Next, I will draw the bottom line.
- Finally, I will add an arc with a radius using the Start, End, and Mid methods.

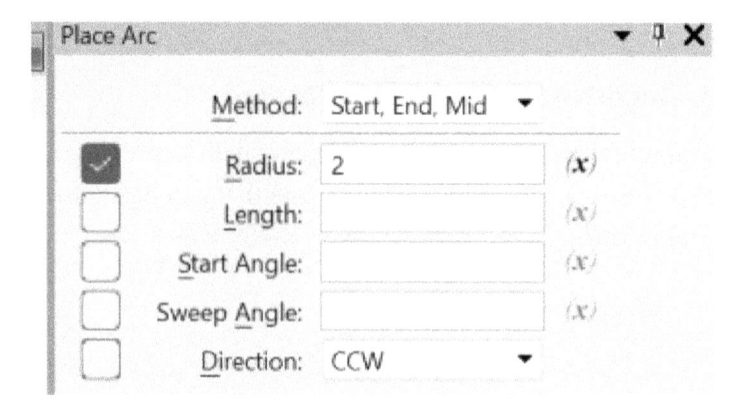

Figure 24 - Place Arc Settings

3. GUIDED DRAWING: Draw a sample shape first to use as a guide. Then, navigate to **Utilities**, click **Record**, and draw the shape. If you don't get it right the first time, stop and start again.

Figure 25 - Record Macro tool

4. FINALIZING THE MACRO: Once done, click **Stop**, enter a name for the macro, and save it as a VBA file by selecting the arrow below. Save it to your computer for further refinement. Test the macro to ensure it performs as intended.

INTRODUCTION TO MICROSTATION VBA 191

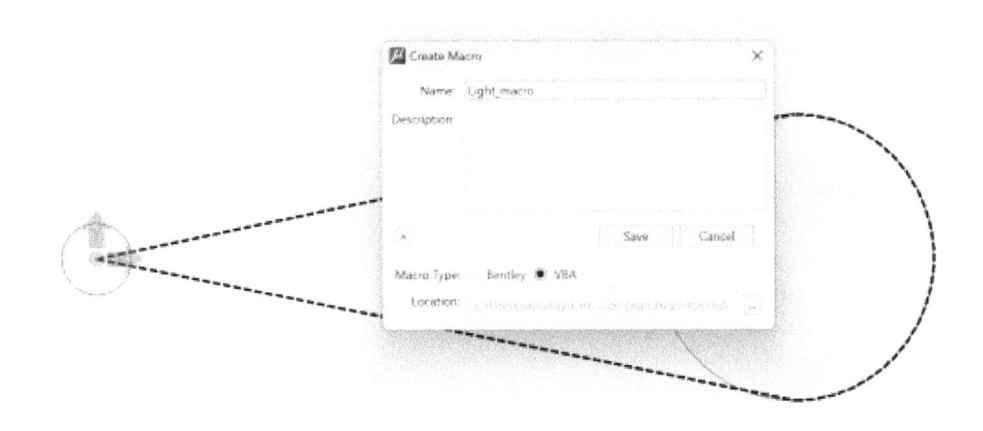

Figure 26 - Saving the Macro

THE CODE SHOULD LOOK like this.

Figure 27 - Sampe of Code

VBA Code 125

```
SUB BMRLIGHT_MACRO()
    Dim startPoint As Point3d
    Dim point As Point3d, point2 As Point3d
```

```
Dim lngTemp As Long
Dim oMessage As CadInputMessage
' Start a command
CadInputQueue.SendCommand "PLACE LINE CONSTRAINED"
' Coordinates are in master units
startPoint.x = 0
startPoint.y = 0
startPoint.z = 0
' Send a data point to the current command
point.x = startPoint.x
point.y = startPoint.y
point.z = startPoint.z
CadInputQueue.SendDataPoint point, 1
point.x = startPoint.x + 10
point.y = startPoint.y + 2
point.z = startPoint.z
CadInputQueue.SendDataPoint point, 1
' Send a reset to the current command
CadInputQueue.SendReset
CadInputQueue.SendCommand "PLACE LINE CONSTRAINED"
point.x = startPoint.x
point.y = startPoint.y
point.z = startPoint.z
CadInputQueue.SendDataPoint point, 1
point.x = startPoint.x + 10
point.y = startPoint.y - 2
point.z = startPoint.z
CadInputQueue.SendDataPoint point, 1
CadInputQueue.SendReset
CadInputQueue.SendCommand "PLACE ARC ICON"
' Set a variable associated with a dialog box
SetCExpressionValue "tcb->msToolSettings.igen.placeArcModeEx", 1, "CONSGEOM"
CadInputQueue.SendCommand "PLACE ARC ICON"
' This only modifies a of the variable it changes. It first
' creates a mask for clearing the bits. Then it gets
' the variable and uses the mask to clear those bits. Finally
' it sets the desired bits in and saves the updated value.
lngTemp = Not 8
lngTemp = GetCExpressionValue _
("lockMask", "CONSGEOM") And lngTemp
SetCExpressionValue "lockMask", lngTemp Or 8, "CONSGEOM"
CadInputQueue.SendCommand _
```

```
"CONSGEOM CONSTRAIN RADIUS UPDATELOCK"
SetCExpressionValue "RadiusValue", "2", "CONSGEOM"
CadInputQueue.SendCommand _
"CONSGEOM CONSTRAIN RADIUS UPDATEVALUE"
point.x = startPoint.x + 10
point.y = startPoint.y - 2
point.z = startPoint.z
CadInputQueue.SendDataPoint point, 1
point.x = startPoint.x + 12
point.y = startPoint.y
point.z = startPoint.z
CadInputQueue.SendDataPoint point, 1
point.x = startPoint.x + 10
point.y = startPoint.y + 2
point.z = startPoint.z
CadInputQueue.SendDataPoint point, 1
CommandState.StartDefaultCommand
End Sub
```

Transitioning to a Function

NOW, I WILL MODIFY the code to create a function that accepts x coordinate, y coordinate, distance, and width. This will be accompanied by a user form.

VBA Code 126

```
SUB MAIN_PROCEDURE_Macro()
    ' Calling the light_shape
    light_shape 0, 0, 20, 4
    End Sub
    Function light_shape(x_coord As Double, _
    y_coord As Double, Distance As Double, width As Double)
    Dim startPoint As Point3d
    Dim point As Point3d, point2 As Point3d
```

• • • •

```
' START A COMMAND
    CadInputQueue.SendCommand "PLACE LINE CONSTRAINED"
    ' Coordinates are in master units
```

```
startPoint.x = x_coord
startPoint.y = y_coord
startPoint.z = 0
' Send a data point to the current command
point.x = startPoint.x
point.y = startPoint.y
point.z = startPoint.z
CadInputQueue.SendDataPoint point, 1
'calculations
x_direction = Distance - width
'top and bottom cone
'vector_x_direction = _
Sqr((cone_size) ^ 2 + (0.5 * (width)) ^ 2)
vector_x_direction = Distance - (0.5 * width)
vector_y_directin = 0.5 * width
```

. . . .

```
POINT.X = STARTPOINT.x + vector_x_direction
    point.y = startPoint.y + vector_y_directin
    point.z = startPoint.z
    CadInputQueue.SendDataPoint point, 1
    ' Send a reset to the current command
    CadInputQueue.SendReset
    CadInputQueue.SendCommand "PLACE LINE CONSTRAINED"
    point.x = startPoint.x
    point.y = startPoint.y
    point.z = startPoint.z
    CadInputQueue.SendDataPoint point, 1
    point.x = startPoint.x + vector_x_direction
    point.y = startPoint.y - vector_y_directin
    point.z = startPoint.z
    CadInputQueue.SendDataPoint point, 1
    CadInputQueue.SendReset
    CadInputQueue.SendCommand "PLACE ARC ICON"
    ' Set a variable associated with a dialog box
    SetCExpressionValue _
    "tcb->msToolSettings.igen.placeArcModeEx", 1, "CONSGEOM"
    CadInputQueue.SendCommand "PLACE ARC ICON"
    lngTemp = Not 8
    lngTemp = GetCExpressionValue _
    ("lockMask", "CONSGEOM") And lngTemp
```

```
SetCExpressionValue "lockMask", lngTemp Or 8, "CONSGEOM"
CadInputQueue.SendCommand _
"CONSGEOM CONSTRAIN RADIUS UPDATELOCK"
SetCExpressionValue "RadiusValue", "2", "CONSGEOM"
CadInputQueue.SendCommand _
"CONSGEOM CONSTRAIN RADIUS UPDATEVALUE"
point.x = startPoint.x + vector_x_direction
point.y = startPoint.y - vector_y_directin
point.z = startPoint.z
CadInputQueue.SendDataPoint point, 1
point.x = startPoint.x + Distance
point.y = startPoint.y
point.z = startPoint.z
CadInputQueue.SendDataPoint point, 1
point.x = startPoint.x + vector_x_direction
point.y = startPoint.y + vector_y_directin
point.z = startPoint.z
CadInputQueue.SendDataPoint point, 1
CommandState.StartDefaultCommand
End Function
```

. . . .

```
SUB SHOWLIGHTZONEFORM()
    frmlightapp.Show modless
    End Sub
```

The form is shown below.

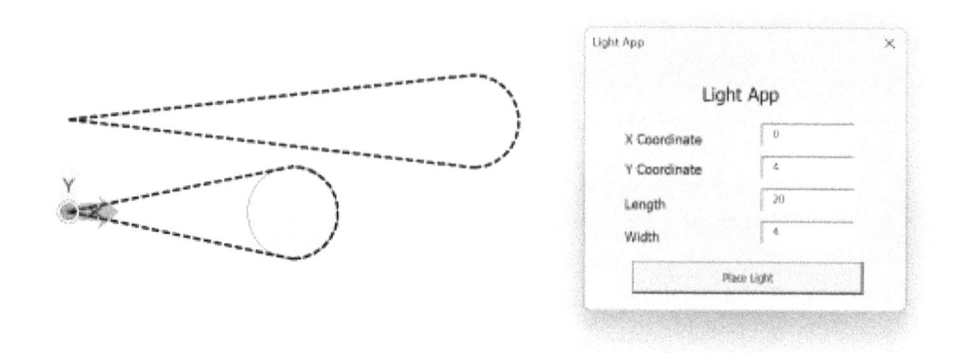

Figure 28 - Macro User Form & Shapes

THE CODE IN THE FORM button.

VBA Code 127

```
PRIVATE BTN_PLACE_LIGHT_Click()
    macro.light_shape txt_x_coord, txt_y_coord, _
    txt_length, txt_width
    End Sub
```

Then Run **ShowLightZoneForm**, this should open the Form, and you can try different values.

Common Use Cases for Macro Recording

- Automating repetitive drawing tasks such as creating standard details or layouts.
- Standardizing workflows for placing cells or elements in consistent locations.
- Streamlining data entry and modification processes across multiple models.

26:
Advanced MicroStation UserForm with Excel Integration

Introduction

In this chapter, we will extend the functionality of our user form to allow not only the editing of existing data but also the creation of new cell directly from the user interface. The created cells will be managed within the MicroStation model and saved in an Excel spreadsheet for future reference and data management.

Objectives

- Create a user form with a list box to display light data.
- Enable users to create new lights through the user form.
- Save both new and edited light data in an Excel file.
- Update the MicroStation model with new lights.

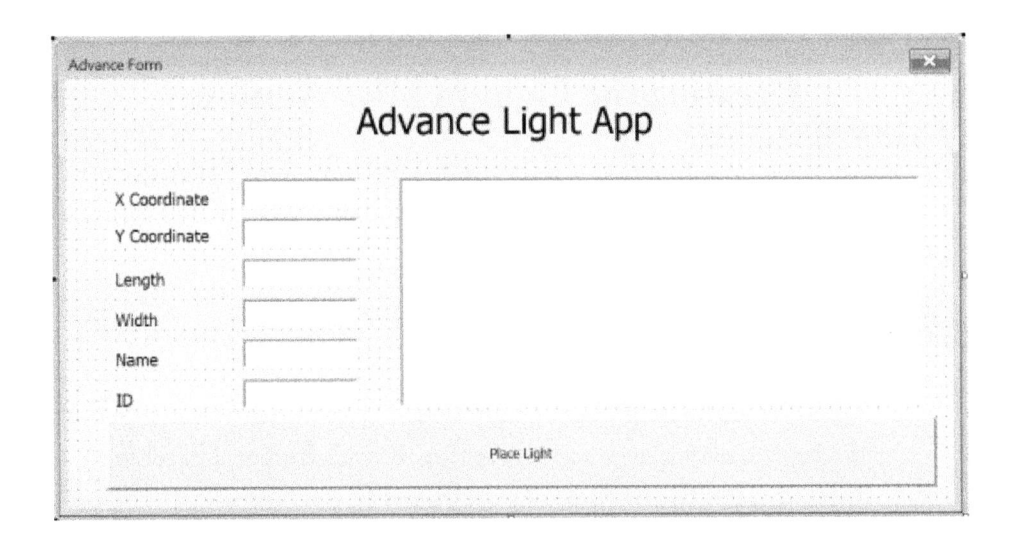

Figure 29 - Advanced UserForm

Creating the UserForm

Designing the UserForm:

- Open the VBA editor in MicroStation.
- Insert a new UserForm.
- Add a ListBox control to display the lights.
- Include TextBox controls for entering light parameters (e.g., position, distance, width).
- Add buttons for actions such as Place Light.

Creating a New Light

TO CREATE A NEW LIGHT from the user form and save its data to Excel, use the following code snippets:

1. Placing a New Light:

- Use the btnPlaceLight_Click subroutine to initiate the creation of a new light. This subroutine captures the input values from the text boxes and calls the light_shape function to create the light.

2. Creating the Light in MicroStation:

- The light_shape function is responsible for defining the light's shape based on the provided parameters. It calculates necessary coordinates and issues commands to draw the light.

3. Loading and Saving Data

- The code for loading existing light data into the ListBox is handled in the PopulateListBoxFromExcel subroutine. It reads from the Excel sheet and populates the ListBox, allowing users to view and select existing lights.

Complete Code Summary

Module Code

```vb
SUB SHOWLIGHTZONEFORM()
    frmAdvance.Show modless
    End Sub
    Sub Main_Procedure_Macro()
    Dim cell_id As Variant ' To hold the retrieved cell ID
    Dim cell_name As String ' To hold the cell name
    ' Set the cell name (ensure this is appropriate for your context)
    cell_name = "P01" ' Example cell name
    ' Calling the light_shape with appropriate parameters
    light_shape 0, 20, 20, 4, cell_name
    End Sub
    Function light_shape(x_coord As Double, y_coord As Double, _
    Distance As Double, width As Double, cell_name As String)
    Dim startPoint As Point3d
    Dim point As Point3d
    Dim vector_x_direction As Double
    Dim vector_y_direction As Double
    Dim cell_id As Variant ' To hold the retrieved cell ID
    ' Start a command
    CadInputQueue.SendCommand "PLACE LINE CONSTRAINED"
    ' Coordinates are in master units
    startPoint.x = x_coord
    startPoint.y = y_coord
    startPoint.z = 0
    ' Send a data point to the current command
    point.x = startPoint.x
    point.y = startPoint.y
    point.z = startPoint.z
    CadInputQueue.SendDataPoint point, 1
    ' Calculations
    vector_x_direction = Distance - (0.5 * width)
```

```
vector_y_direction = 0.5 * width
' Define the points for the shape
point.x = startPoint.x + vector_x_direction
point.y = startPoint.y + vector_y_direction
point.z = startPoint.z
CadInputQueue.SendDataPoint point, 1
' Send a reset to the current command
CadInputQueue.SendReset
CadInputQueue.SendCommand "PLACE LINE CONSTRAINED"
point.x = startPoint.x
point.y = startPoint.y
point.z = startPoint.z
CadInputQueue.SendDataPoint point, 1
point.x = startPoint.x + vector_x_direction
point.y = startPoint.y - vector_y_direction
point.z = startPoint.z
CadInputQueue.SendDataPoint point, 1
CadInputQueue.SendReset
CadInputQueue.SendCommand "PLACE ARC ICON"
' Set a variable associated with a dialog box
    SetCExpressionValue    "tcb->msToolSettings.igen.placeArcModeEx",    1,
"CONSGEOM"
    CadInputQueue.SendCommand "PLACE ARC ICON"
' Modify the variable for the arc placement
Dim lngTemp As Long
lngTemp = Not 8
    lngTemp  =  GetCExpressionValue("lockMask",  "CONSGEOM")  And
lngTemp
    SetCExpressionValue "lockMask", lngTemp Or 8, "CONSGEOM"
    CadInputQueue.SendCommand    "CONSGEOM    CONSTRAIN
RADIUS UPDATELOCK"
    SetCExpressionValue "RadiusValue", "2", "CONSGEOM"
    CadInputQueue.SendCommand    "CONSGEOM    CONSTRAIN
RADIUS UPDATEVALUE"
    ' Define the remaining points of the shape
```

```vb
    point.x = startPoint.x + vector_x_direction
    point.y = startPoint.y - vector_y_direction
    point.z = startPoint.z
    CadInputQueue.SendDataPoint point, 1
    point.x = startPoint.x + Distance
    point.y = startPoint.y
    point.z = startPoint.z
    CadInputQueue.SendDataPoint point, 1
    point.x = startPoint.x + vector_x_direction
    point.y = startPoint.y + vector_y_direction
    point.z = startPoint.z
    CadInputQueue.SendDataPoint point, 1
    CommandState.StartDefaultCommand
    ' Retrieve and group unlocked visible elements
    RetrieveAndGroupUnlockedVisibleElements
    ' Read cell ID
    cell_id = RetrieveElementID()
    If IsNumeric(cell_id) Then
    ' Cell rename
    RenameCellByID CLng(cell_id), cell_name ' Convert to Long before passing
    ' Save to Excel
    Call Save_to_Excel(CLng(cell_id), cell_name, x_coord, y_coord, Distance, width)
    Else
    Debug.Print "Error: " & cell_id ' Handle the case where no valid ID was found
    End If
    End Function
    Sub RetrieveAndGroupUnlockedVisibleElements()
    Dim oElement As element
    Dim oEnum As elementEnumerator
    Dim MyScanCriteria As ElementScanCriteria
    Dim elementID As Variant
    Dim elementIDs As Collection
```

```vba
Set elementIDs = New Collection
' Set scan criteria to include only visible elements
Set MyScanCriteria = New ElementScanCriteria
MyScanCriteria.IncludeOnlyVisible
' Scan the active model
Set oEnum = ActiveModelReference.Scan(MyScanCriteria)
' Loop through the elements found by the scan
Do While oEnum.MoveNext
Set oElement = oEnum.Current
' Check if the element is valid and unlocked
If Not oElement Is Nothing Then
If Not oElement.IsLocked Then ' Check if the element is unlocked
' Safely retrieve the element ID64
On Error Resume Next
elementID = oElement.ID64 ' Retrieve the 64-bit element ID
On Error GoTo 0
' Add the ID to the collection if valid
If elementID <> Empty Then
Debug.Print "Element ID64: " & elementID
elementIDs.Add elementID ' Store the element ID
Else
Debug.Print "Could not retrieve ID64 for this element."
End If
End If
End If
Loop
' Check if any IDs were found and group them
If elementIDs.Count > 0 Then
Dim i As Long
Dim oElementToGroup As element
' Select and group each element by ID
For i = 1 To elementIDs.Count
Set oElementToGroup = _
ActiveModelReference.GetElementByID64(elementIDs(i))
If Not oElementToGroup Is Nothing Then
```

```
    ActiveModelReference.SelectElement oElementToGroup, True
    End If
Next i
' Execute the GROUP command
CadInputQueue.SendCommand "GROUP"
' Unselect all elements after grouping
ActiveModelReference.UnselectAllElements
CommandState.StartDefaultCommand
    Debug.Print "Grouped " & elementIDs.Count & " unlocked visible elements."
Else
    Debug.Print "No unlocked visible elements found to group."
End If
End Sub
Function RetrieveElementID() As Variant
Dim oElement As element
Dim oEnum As elementEnumerator
Dim elementID As Variant
' Use Variant to handle potential large values from ID64
' Set scan criteria if needed (optional)
Dim MyScanCriteria As ElementScanCriteria
Set MyScanCriteria = New ElementScanCriteria
MyScanCriteria.IncludeOnlyVisible
' Optional: include only visible elements
' Scan the active model
Set oEnum = ActiveModelReference.Scan(MyScanCriteria)
' Loop through the elements found by the scan
Do While oEnum.MoveNext
    Set oElement = oEnum.Current
    ' Check if the element is valid
    If Not oElement Is Nothing Then
    ' Safely retrieve the element ID64
    On Error Resume Next
    elementID = oElement.ID64 ' Retrieve the 64-bit element ID
    On Error GoTo 0
```

```vba
' Print the ID in the Immediate Window if no error
If elementID <> Empty Then
RetrieveElementID = elementID
Debug.Print "Element ID64: " & elementID
Else
Debug.Print "Could not retrieve ID64 for this element."
End If
End If
Loop
End Function
Function RenameCellByID(elementID As Long, newCellName As String)
    Dim oCell As CellElement
    ' Retrieve the cell element by its ID
    Set oCell = ActiveModelReference.GetElementByID64(elementID)
    ' Rename the cell
    oCell.Name = newCellName
    ' Commit the changes to the model
    oCell.Rewrite
    Debug.Print "New name Given.."
End Function
Sub Save_to_Excel(cell_id As Double, cell_name As String, _
x_coord As Double, y_coord As Double, Distance As Double, width As Double)
    Dim xlApp As Excel.Application
    Dim wb As Excel.Workbook
    Dim ws As Excel.Worksheet
    Dim row As Long
    Dim lastRow As Long
    ' Get the active Excel application and workbook
    Set xlApp = GetObject(, "Excel.Application")
    Set wb = xlApp.ActiveWorkbook
    Set ws = wb.ActiveSheet ' Use the active sheet directly
    ' Find the last row with data in the active sheet
    lastRow = ws.Cells(ws.Rows.Count, 1).End(xlUp).Row
```

```vb
' Write the headers to the Excel sheet if the first row is empty
If lastRow = 1 Then
ws.Cells(1, 1).Value = "Element ID"
ws.Cells(1, 2).Value = "Name"
ws.Cells(1, 3).Value = "X Coordinate"
ws.Cells(1, 4).Value = "Y Coordinate"
ws.Cells(1, 5).Value = "Length"
ws.Cells(1, 6).Value = "Width"
' Format the header row
With ws.Range(ws.Cells(1, 1), ws.Cells(1, 6))
.Interior.Color = RGB(0, 255, 0) ' Green background
.Font.Bold = True ' Bold font
End With
End If
' Write data to the next row
row = lastRow + 1
ws.Cells(row, 1).Value = cell_id
ws.Cells(row, 2).Value = cell_name
ws.Cells(row, 3).Value = x_coord
ws.Cells(row, 4).Value = y_coord
ws.Cells(row, 5).Value = Distance
ws.Cells(row, 6).Value = width ' Adjust as necessary
' Clear the existing items in the list box before adding new ones
frmAdvance.lblight.Clear
' Set the list box to have multiple columns
frmAdvance.lblight.ColumnCount = 6 ' Number of columns in the list box
frmAdvance.lblight.ColumnWidths = "50;50;50;50;50;50" ' Adjust widths as needed
    ' Add headings to the list box
    Dim headings As Variant
    headings = Array("Element ID", "Cell Name", "X Coordinate", _
    "Y Coordinate", "Length", "Width")
    ' Add the headings as the first row
    frmAdvance.lblight.AddItem
    For i = LBound(headings) To UBound(headings)
```

```vb
frmAdvance.lblight.List(0, i) = headings(i)
Next i
' Lock the element
LockElement CLng(cell_id) ' Ensure the cell_id is passed as a Long
' Populate the list box with the new values
Call PopulateListBoxFromExcel
End Sub
Sub PopulateListBoxFromExcel()
Dim xlApp As Excel.Application
Dim wb As Excel.Workbook
Dim ws As Excel.Worksheet
Dim lastRow As Long
Dim i As Long
' Get the active Excel application and workbook
Set xlApp = GetObject(, "Excel.Application")
Set wb = xlApp.ActiveWorkbook
Set ws = wb.ActiveSheet ' Use the active sheet directly
' Get the last row to know the number of items to read
lastRow = ws.Cells(ws.Rows.Count, 1).End(xlUp).Row
' Loop through rows and add items to the list box
For i = 2 To lastRow ' Start from 2 to skip the header
frmAdvance.lblight.AddItem
frmAdvance.lblight.List(frmAdvance.lblight.ListCount - 1, 0) = _
ws.Cells(i, 1).Value ' Element ID
frmAdvance.lblight.List(frmAdvance.lblight.ListCount - 1, 1) = _
ws.Cells(i, 2).Value ' Cell Name
frmAdvance.lblight.List(frmAdvance.lblight.ListCount - 1, 2) = _
ws.Cells(i, 3).Value ' X Coordinate
frmAdvance.lblight.List(frmAdvance.lblight.ListCount - 1, 3) = _
ws.Cells(i, 4).Value ' Y Coordinate
frmAdvance.lblight.List(frmAdvance.lblight.ListCount - 1, 4) = _
ws.Cells(i, 5).Value ' Length
frmAdvance.lblight.List(frmAdvance.lblight.ListCount - 1, 5) = _
ws.Cells(i, 6).Value ' Width
Next i
```

```vba
End Sub
Sub SelectElementByID(elementID As Long)
Dim oElement As element
ActiveModelReference.UnselectAllElements
' Retrieve the element by ID
Set oElement = ActiveModelReference.GetElementByID64(elementID)
' Select the element if it exists
If Not oElement Is Nothing Then
ActiveModelReference.SelectElement oElement, True
Debug.Print "Element with ID " & elementID & " selected."
Else
Debug.Print "Element with ID " & elementID & " not found."
End If
End Sub
Sub LockElement(elementID As Long)
Dim oElement As element
ActiveModelReference.UnselectAllElements
' Retrieve the element by ID
Set oElement = ActiveModelReference.GetElementByID64(elementID)
' Lock the element
oElement.IsLocked = True
' Write the changes back to the model
oElement.Rewrite
End Sub
```

Form Code

```vba
PRIVATE SUB BTNPLACELIGHT_Click()
    Dim cell_id As Variant ' To hold the retrieved cell ID
    Dim cell_name As String ' To hold the cell name
    ' Set the cell name based on user input
    cell_name = txtname ' Example cell name
    ' Calling the light_shape with appropriate parameters
    zb_advance_form.light_shape txt_x_coord, txt_y_coord, txt_length, _
    txt_width, cell_name
```

```vba
End Sub
Private Sub lblight_DblClick(ByVal Cancel As MSForms.ReturnBoolean)
Dim selectedRow As Long
' Get the selected row index
selectedRow = lblight.ListIndex
' Check if an item is selected
If selectedRow >= 0 Then ' Check if the index is valid (greater than or equal to 0)
    ' Debug: Print the selected row index
    Debug.Print "Selected Row: " & selectedRow
    ' Populate text boxes with the selected row values
    txtElementID.Value = lblight.List(selectedRow, 0) ' Element ID
    txtname.Value = lblight.List(selectedRow, 1) ' Cell Name
    txt_x_coord.Value = lblight.List(selectedRow, 2) ' X Coordinate
    txt_y_coord.Value = lblight.List(selectedRow, 3) ' Y Coordinate
    txt_length.Value = lblight.List(selectedRow, 4) ' Length
    txt_width.Value = lblight.List(selectedRow, 5) ' Width
Else
    Debug.Print "No item selected in the list box."
End If
zb_advance_form.SelectElementByID txtElementID
End Sub
Private Sub UserForm_Initialize()
' Clear existing items
lblight.Clear
' Set the list box to have multiple columns
lblight.ColumnCount = 6 ' Number of columns in the list box
lblight.ColumnWidths = "50;50;50;50;50;50" ' Adjust widths as needed
' Add headings to the list box
Dim headings As Variant
headings = Array("Element ID", "Cell Name", "X Coordinate", _
"Y Coordinate", "Length", "Width")
' Add the headings as the first row
lblight.AddItem
For i = LBound(headings) To UBound(headings)
```

INTRODUCTION TO MICROSTATION VBA 209

lblight.List(0, i) = headings(i)
Next i
' Populate the list box with existing data from Excel
Call zb_advance_form.PopulateListBoxFromExcel
End Sub

This completes the chapter on the Advanced MicroStation UserForm with Light Creation and Excel Integration.

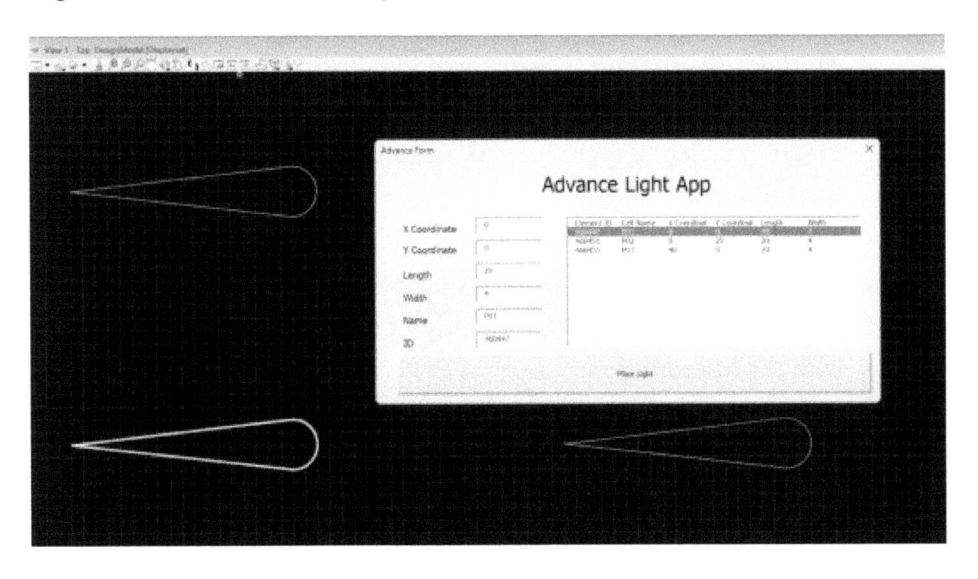

Figure 30 - Advance VBA UserForm

Element ID	Name	X Coordinate	Y Coordinate	Length	Width
468447	P01	0	0	20	4
468451	P02	0	20	20	4
468455	P03	40	0	20	4

Figure 31 - Excel Table

Conclusion

BY FOLLOWING THE STEPS outlined in this chapter, you can create a user form in MicroStation that allows users to easily create and edit lights. The integration with Excel ensures that all light data is saved and retrievable, enhancing project management and collaboration.

Reasoning Behind the Approach

- **Modularity**: The code is structured into distinct functions and subroutines, promoting reusability and ease of maintenance. Each function performs a specific task, simplifying debugging and enhancements.
- **User Interaction**: The user form provides a graphical interface that facilitates user interaction with the application. Users can quickly create, view, and modify lights without needing to interact directly with the Microstation.
- **Data Management**: Using Excel as a backend for storing light placements allows for easy data management and reporting. Users can generate reports or perform further analyses based on the saved data.

27. Creating a Micro Station Add-on

Developing Custom Add-ons with VBA

VBA (Visual Basic for Applications) empowers users to create fully-fledged add-ons for MicroStation, enhancing functionality and streamlining workflows. These custom tools can significantly improve efficiency in handling specific tasks or processes that are not natively supported by MicroStation. By creating add-ons, users can tailor the software to better fit their individual needs and project requirements.

Steps to Create a Simple Add-on

1. OPEN YOUR GUI DGNLIB File:

- Start by launching MicroStation CONNECT and opening your GUI DGNlib file, which contains the user interface components you plan to customize. The GUI DGNlib file is essential as it serves as a repository for your custom toolboxes and dialogs.

2. Access the Customize Dialog:

- Navigate to the menu: File > Settings > Configuration > Customize. This opens the Customize dialog, where you can modify toolbars, menus, and other interface elements.

3. Create Custom Toolbox:

- If your custom toolbox is not already placed in a User Task, Create new toolbox and create a new tool after that. This allows you to access your toolbox more efficiently from the task navigation pane. Or right-click on the toolbox in the dialog and select the appropriate option to copy it.

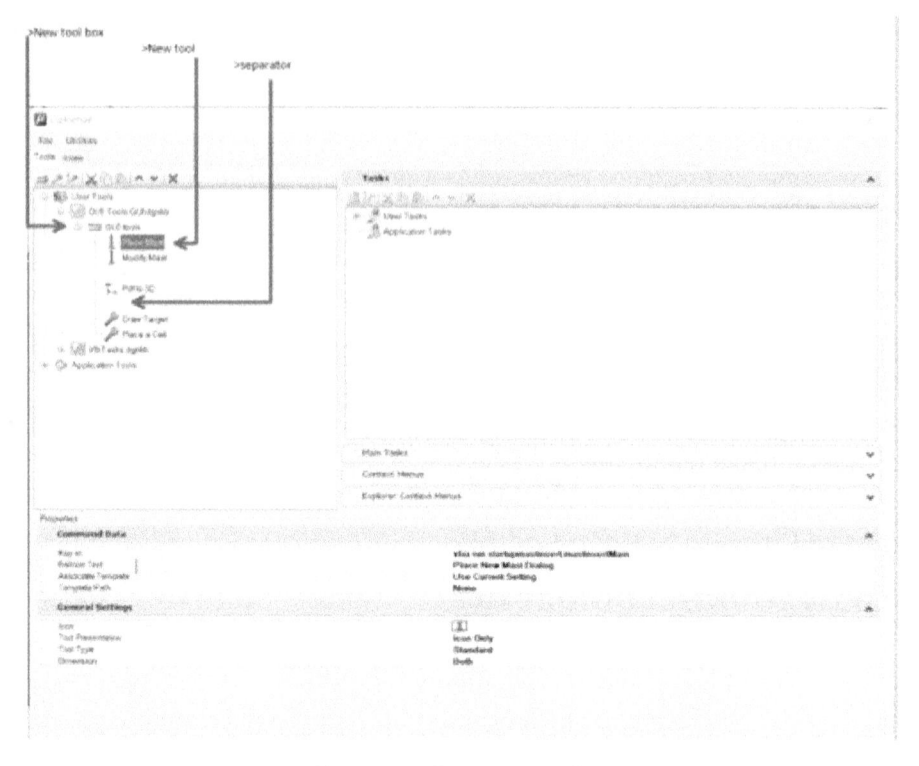

Figure 32 - Customise window

1. **Add Module Name and Subroutine Name**:

   - In the Key-in box of the Customize dialog, enter the module name and the specific subroutine you wish to execute when the tool is activated. The format typically looks like this: VBA RUN [ModuleName].[SubroutineName]. This links your custom VBA functionality to the button or tool you are creating.

INTRODUCTION TO MICROSTATION VBA

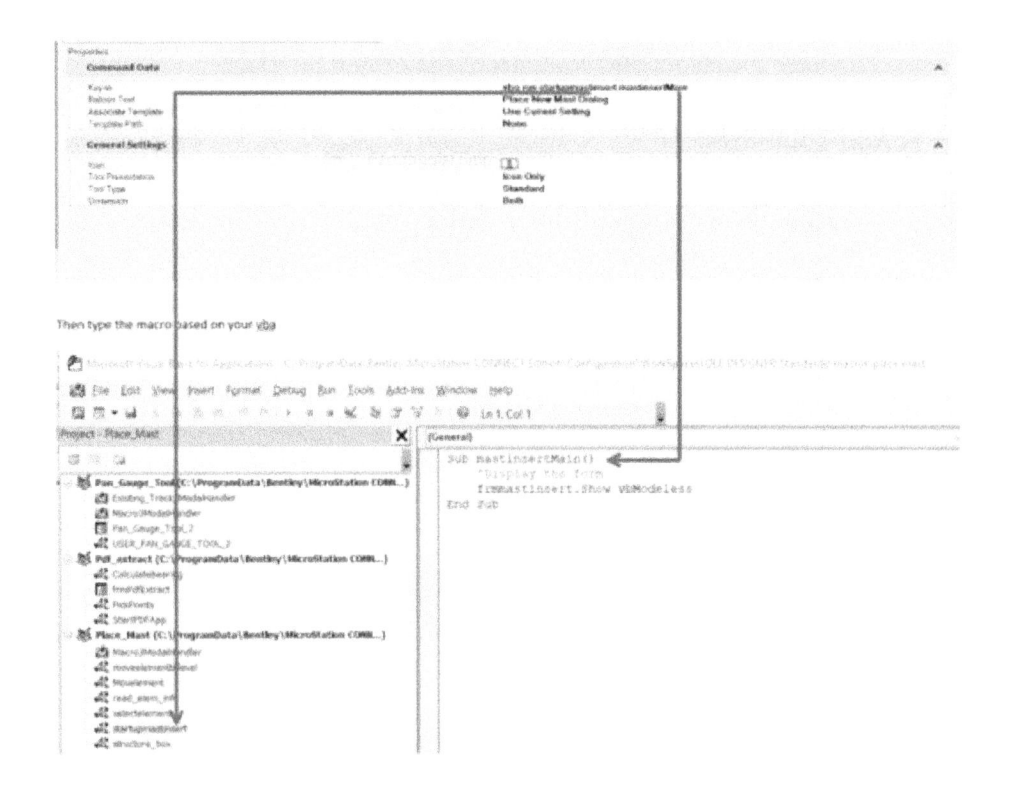

Figure 33 - Adding Module name

5. **Close the Customize Dialog**:

    - After entering your module and subroutine information, close the Customize dialog. This action saves your changes and updates the interface.

6. **Select Task Navigation**:

    - In the Active Workflow drop-down menu, select Task Navigation. This will allow you to modify the task interface to include your custom tools.

7. **Customize the Ribbon**:

    - Right-click on the ribbon within the Task Selection area and choose

Customize Ribbon. This will open a new dialog where you can manage the layout and contents of the ribbon interface.

8. **Add Toolboxes**:

- Within the Customize Ribbon dialog, look for the "Toolboxes (Custom)" section. This displays the GUI applications you have on the left side of the dialog. On the right side, you can create a new Workflow and group or button to migrate your tools.

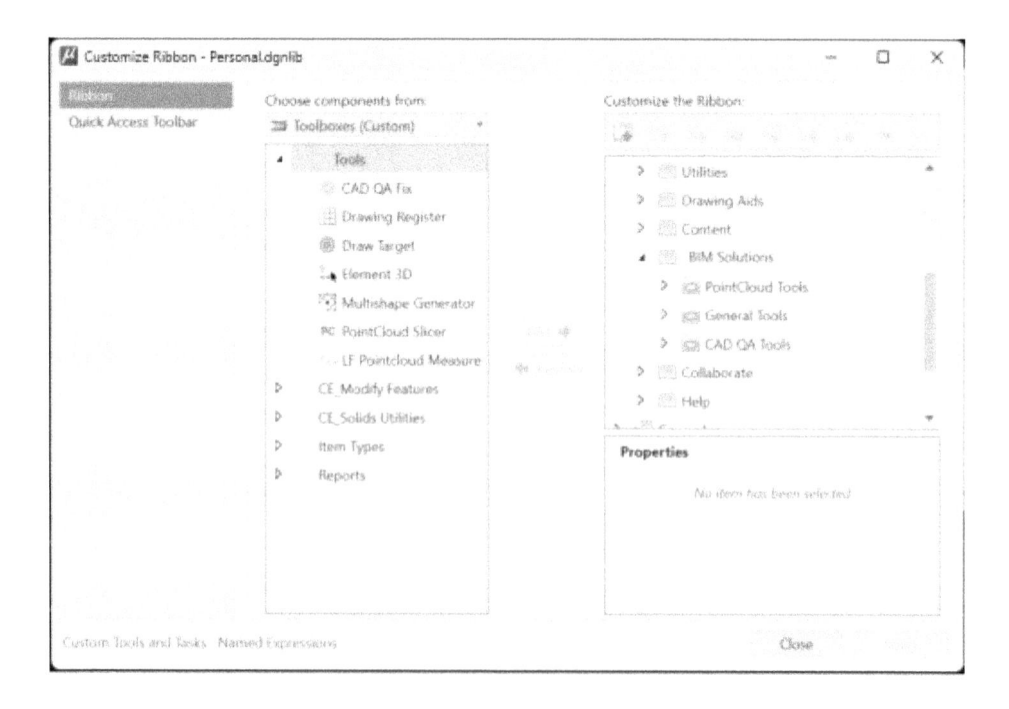

Figure 34 - Customise Ribbon window

9. **Select the VBA Component**:

- From the available toolboxes, select your VBA component and click the "Add" button. This action will place your custom icon into the Workflow, allowing easy access from the ribbon. Remember to adjust the position where you want the icon to appear.

10. **Apply Changes and Close the Customize Ribbon Window**:

- Once you've arranged your components, click "Apply" and then "Close" to finalize the changes. Your custom toolbox will now be displayed in the ribbon for easy access during your MicroStation sessions.

Your custom toolbox will be displayed in the ribbon

Best Practices for Add-ons

- **User-Centric Design**: Ensure that the add-on is intuitive and user-friendly. Conduct user testing or gather feedback from potential users to identify areas for improvement before the final release. A well-designed interface can significantly enhance user adoption and satisfaction.
- **Thorough Documentation**: Document the add-on's functionality, installation steps, and user instructions. Clear documentation helps users understand how to leverage the add-on effectively and troubleshoot any issues they may encounter.
- **Version Control**: Maintain a version control system for your add-on. This practice allows for easy tracking of changes and updates over time, facilitating collaboration among multiple developers and ensuring that users have access to the latest features and fixes.

EXAMPLES OF POTENTIAL Add-ons

- **Batch Processing Tool**: Automate modifications of multiple elements across several files, saving time and effort for tasks that would otherwise require manual intervention.
- **Custom Reporting Tool**: Generate summaries of model data, providing improved oversight and insights into project progress. Such a tool could enable users to export reports directly to Excel or PDF formats.
- **Interactive Design Checker**: Validate models against predefined

standards, ensuring compliance with project requirements and improving overall design quality.

Closing Summary

As we reach the end of this book, you should now have a solid understanding of how to use VBA to automate and enhance your work within MicroStation. You have learned the fundamentals of the VBA language and its integrated development environment (IDE), as well as how to create complex applications like the Cells App and Light Shape Generator. This journey has equipped you with the skills to significantly improve your efficiency and workflow.

The projects, exercises, and examples in this book have demonstrated the versatility of VBA, showing how it can be applied to automate tasks, manipulate elements, interact with external applications like Excel, and even create custom tools and add-ons. Whether you are simplifying routine tasks or tackling more advanced design challenges, you now have the knowledge to develop robust VBA solutions tailored to your needs.

Remember, VBA is a powerful tool that grows with you. The more you experiment, the more you will uncover its potential. As MicroStation evolves, so too will the possibilities with VBA, allowing you to continue improving your workflows and optimizing your design processes.

Thank you for taking this journey into MicroStation VBA automation. We hope this book has provided you with valuable insights, practical skills, and the inspiration to keep innovating in your design work. As you continue to explore and apply what you've learned, there is no limit to the efficiencies and custom solutions you can create. Happy automating!

About the Author

Saeed Murray is a distinguished civil engineer with a First-Class degree from a prestigious Scottish university. With over 15 years of experience in the engineering field, Saeed has become an expert in Building Information Modelling (BIM) and automation. His work spans various industries, including civil engineering, railways, and telecommunications, where he has consistently leveraged his expertise to automate BIM and CAD processes through the use of VBA.

Saeed's book on VBA is the culmination of extensive research into best practices and innovative solutions developed throughout his career. His approach has not only streamlined workflows but has also resulted in substantial time and cost savings for numerous projects. Passionate about improving efficiency in engineering, Saeed has dedicated his career to creating tailored solutions that benefit both projects and businesses alike.

If you encounter any issues with the code or wish to access the source code, you can download it at:

https://gitfront.io/r/smurraygit/a7RWUfnpmHTW/Microstation-VBA-Introduction/

Milton Keynes UK
Ingram Content Group UK Ltd.
UKHW020719101024
449496UK00010B/233